how to know the grasses

The **Pictured Key Nature Series** has been published since 1944 by the Wm. C. Brown Company. The series was initiated in 1937 by the late Dr. H. E. Jaques, Professor Emeritus of Biology at Iowa Wesleyan University. Dr. Jaques' dedication to the interest of nature lovers in every walk of life has resulted in the prominent place this series fills for all who wonder **"How to Know."**

John F. Bamrick and Edward T. Cawley
Consulting Editors

Louise K. Barrett
Editor

The Pictured Key Nature Series

How to Know the

how
to
know
the
grasses

Third Edition

Richard W. Pohl
Iowa State University

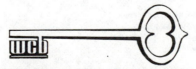

The Pictured Key Nature Series
Wm. C. Brown Company Publishers
Dubuque, Iowa

Contents

Preface

This edition of *How to Know the Grasses* includes keys and illustrations for 324 of the most common and important American grasses—those that the beginner is most likely to encounter, including those of importance in farming, gardening, weed control, and range and pasture management. Some less common or rare grasses have been included because of their peculiar structure or interesting evolutionary relationships. In addition to those keyed and illustrated, 124 others are mentioned in connection with closely related species, and their distinguishing features are pointed out.

The system of classification used in this book is based largely on one first proposed by G. Ledyard Stebbins, Jr. and Beecher Crampton. This system, first discussed at an International Botanical Congress in 1959, is based upon many characteristics of the vegetative and reproductive parts of the plants, in addition to the spikelet and inflorescence characters used in the classic system of A. S. Hitchcock. The new system is based on such features as the number and size of the chromosomes, the types of specialized cells in the leaf epidermis, the arrangement of cells in the leaf cross section, the nature of the embryo and the young seedling, the type of stored carbohydrates, and other traits. The beginner may wonder why we bother with such fine details, since they cannot be observed readily with the hand lens or low power microscope. Unfortunately, the external traits of many grasses are so similar, and so subject to convergent evolution, that they have not proved to be reliable indicators of relationships in many cases. The use of the new types of characteristics has resulted in the formulation of a new system of classification that includes about six subfamilies instead of the two that Hitchcock used. This system is a remarkable improvement over the older one in many respects, and has many theoretical and practical benefits. For example, it is now possible, using nothing more than a small bit of a grass leaf blade, to determine the subfamily by making a crude cross section with a razor blade, and observing this and a bit of the epidermis with the compound microscope. Many puzzles of relationships can be resolved quickly in this fashion, even when the spikelet characters are inconclusive or misleading.

Since the new system places such heavy reliance on the use of microscopic traits that are not readily observable to many persons, I have had to make new "artificial" keys to the genera, without considering the subfamilies, based upon the most easily seen characteristics of the plants. The main key has been subdivided into six shorter keys, to made identification easier.

The naming of genera and species in the book basically follows that of the second edition of Hitchcock's Manual of the Grasses of the United States, with alterations needed to bring the names in accord with recent revisions. A few new generic names have been introduced, and one has been abandoned.

The majority of the illustrations are the work of the author. Others, marked with the initial B, are the work of Mrs. John Bardach, whose assistance is gratefully acknowledged. A number of the others were drawn by Elsie H. Froeschner and appeared originally in the author's *Grasses of Iowa*. They are used here by permission of the Iowa State University Journal of Research.

Lachryma Iob

Simler
1566 AD

An early illustration of a grass, Job's tears, *Coix lacryma-jobi*

What Is a Grass?

Of all the world's flowering plants, the grasses are undoubtedly the most important to man. They contribute tremendously to the earth's green mantle of vegetation; they are the source of the principal foods of man and his domestic animals. Without the grasses, agriculture would be virtually impossible: grain, sugar, syrup, spice, paper, perfume, pasture, oil and timber, and a thousand other items of daily use are products of various grasses. They hold the hills, plains and mountains against the destructive erosive forces of wind and water. In the end, they form the sod that covers the sleeping dead.

Despite the fact that the grasses are so important to us, we usually know little about them. Why? Because we think that "All grasses are alike," or "They are too hard to tell apart." But neither statement is true. There are over 5,000 "kinds" or species of grasses in the world and 1,400 of these are found in the United States. This book contains descriptions and pictures of over 400 of the more common grasses of our country. While many are superficially similar, they all have good individual marks of recognition. Nobody would at a second glance, for example, confuse foxtail and corn, or quackgrass and oats, or Sudan grass and barley, yet these are all grasses, members of one natural family, the GRAMINEAE, or grass family.

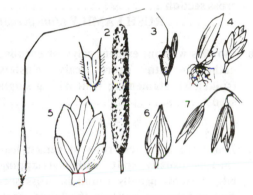

Grasses are easy to recognize. Here are some common ones. 1. Floret of porcupine grass *(Stipa)*. 2. Spikelet and panicle of timothy *(Phleum)*. 3. Branchlet with spikelets of Johnson grass *(Sorghum)*. 4. Floret and spikelet of Kentucky bluegrass *(Poa)*. 5. Spikelet of wheat *(Triticum)*. 6. Spikelet of proso millet *(Panicum)*. 7. Spikelets of oats *(Avena)*.

HOW TO RECOGNIZE THE GRASS FAMILY

The grasses and their allies are all members of the great group of flowering plants which we call the MONOCOTYLEDONS. The members of this group are alike in having one seed leaf, parallel-veined leaves (with few exceptions), and stems in which the vascular bundles are scattered in the pith. Among the monocotyledons, members of three families of plants

1

have a "grasslike" appearance and may be confused. These are the grasses (Gramineae), the sedges (Cyperaceae), and the rushes (Juncaceae). A little study of the following key and pictures will show how to separate them quickly and surely.

KEY TO GRASSES, SEDGES, AND RUSHES

1a Flowers with stiff, greenish or brownish, 6 parted perianth (calyx and corolla); stamens 6 or 3; fruit a many-seeded capsule; leaves usually wiry and round in cross section . . . .
. . . . **RUSH FAMILY** *(Juncaceae)*

1b Flowers without evident calyx or corolla, gathered into short scaly clusters (spikelets); stamens 3; fruit with a single seed **2**

2a Leaves in 2 vertical rows or ranks; leaf sheaths usually split, with overlapping edges; stems usually round in cross section and hollow between the joints; each flower of the spikelet contained between 2 bracts, the lemma and the palea . . . . . .
. . . . **GRASS FAMILY** *(Gramineae)*

2b Leaves in 3 vertical rows or ranks; leaf sheaths tubular, not split; stems often triangular in cross section and solid between joints; each flower of the spikelet in the axil of a single bract, the glume . . . . .
. . . . **SEDGE FAMILY** *(Cyperaceae)*

GRASS   SEDGE   RUSH

Figure 1

What Do Grasses Look Like?

ROOTS

The root systems of grasses (Fig. 2) are always fibrous. While the germinating seed produces a primary root, this soon dies and in most cases is replaced by numerous roots that arise from the basal nodes of the stems. Because of the fibrous nature of the roots, they are excellent soil binders. When we pull up a grass plant, we remove only a small portion of the total root system. In some species, the roots may reach a depth of two meters or more.

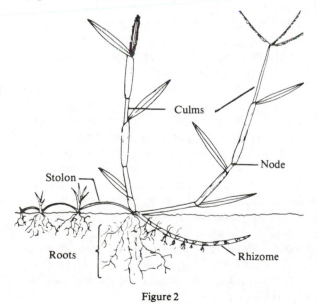

Figure 2

STEMS

The flowering stems of grasses are called culms (Fig. 2). They are jointed and the internodes are usually round and hollow, but the joints (nodes) are solid. They may branch, in which instance a thin membrane, H-shaped in cross section, lies between the main culm and the branch. This is the prophyllum (Fig. 3), the strangely modified first leaf of the branch. It clasps the main culm with two flanges and the branch with the other two. Thus it serves to brace the weak V-joint between the main stem and the branch. Stems may be erect, or with bent, knee-like bases (decumbent), or they may trail on the surface of the ground. Such creeping stems, called stolons, bear foliage leaves. Underground stems, called rhizomes, bear only scale-like leaves and are non-green. Both stolons and rhizomes are important means of reproduction and aid the grasses in coping with grazing. Short, leafy, non-blooming basal stems are called innovations. They add to the density and desirability of such turf grasses as Kentucky bluegrass. The stems of some alpine and desert grasses are only a few centimeters tall, while those of the giant timber bamboos may reach as much as 35 meters in length. Most grass stems are between one and five millimeters thick. Those of the larger bamboos may reach 10-15 centimeters in thickness.

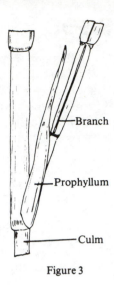

Figure 3

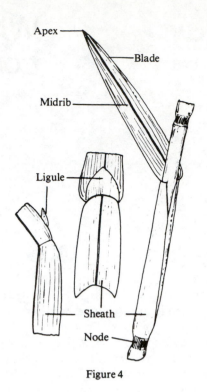

Figure 4

LEAVES

The leaves of grasses almost always have parallel-veined blades that are usually narrow and have parallel sides. The foliage leaf (Fig. 4) has three main parts: the sheath, the ligule, and the blade. There is no true petiole, although in the bamboos a short constriction, the pseudo-petiole, occurs between the sheath and the blade. The sheath is the split tubular portion surrounding the culm. The ligule is a little membranous or hairy collar that stands up at the junction of the sheath and the blade. The elongated blade usually has a conspicuous midrib as well as numerous smaller "nerves" or "veins" parallel to it. The tip, or apex, of the blade is sharp pointed. Little projections of the leaf base or sheath apex are called auricles. In a few genera, notably *Bromus, Dactylis,* and the genera of the tribe Meliceae *(Melica, Schizachne, Glyceria, Pleuropogon),* the leaf sheath has united edges, forming a tube, much as in the sedge family.

FLOWER CLUSTERS

The flower clusters or inflorescences (Fig. 5) of grasses always bear small scaly units called spikelets. These are arranged in panicles, balanced or symmetrical spikes, one-sided spikes (with all the spikelets borne on the lower side of the rachis), racemes, or in specialized aggregations to which I have given the name RAME. A rame is an unbranched inflorescence that is similar to a spike in having some sessile (stalkless) spikelets, and similar to a raceme in also having some stalked or pedicellate spikelets. The rame is the characteristic inflorescence of many members of the Andropogoneae, the bluestem tribe. In this group, each rachis internode may bear one sessile and one stalked spikelet. Rames are also found in the genus *Hordeum,* where each unit of the rachis bears one sessile and two stalked spikelets.

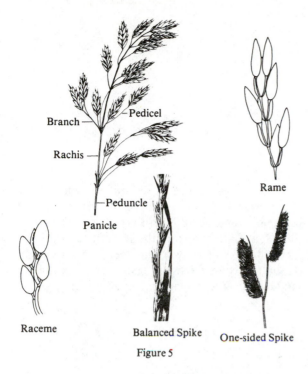

Figure 5

The parts of the inflorescence include the stalk or peduncle, which is continued up into the main axis or rachis of the inflorescence. There may be branches of several ranks. The stalk of the individual spikelet is called a pedicel. At the base of each branch there may be found a little swelling or pulvinus, which helps to spread the branches of the inflorescence apart when they emerge from the uppermost sheath.

SPIKELETS

Since the flowers of grasses are minute, simple, and very similar, they are rarely used in identification. Instead, we look for differences in the bracts (modified leaves) which surround the flowers. The unit subdivisions of the inflorescence are called spikelets (Fig. 6). The simplest sort of spikelet is merely a tiny scaly branchlet of flowers, each flower being surrounded by two bracts. At the base of this branchlet there are two bracts which have no flowers in their axils. These are the first and second glumes. The remainder of the spikelet is made up of flowering units called florets, which are arranged alternately in two rows on a central stalk, called a rachilla, which is usually concealed by the overlapping florets. Each floret consists of an outer bract or lemma and an inner bract or palea, with the naked flower between them. The lemma corresponds to an ordinary foliage leaf, the palea to the prophyllum, and the flower to a branch. During the brief time of flowering, two little blisters, the lodicules, which lie between the ovary of the flower and the lemma, swell up and force the lemma outwards. This allows the stigmas and stamens to protrude. The lodicules are the evolutionary vestiges of a calyx or corolla. If you can find smooth brome grass, orchard grass, or any one of many other grasses in bloom in the morning dew, you can usually observe the lodicules with a hand lens. The actual flower consists of two lodicules, three stamens with long slender filaments, and an ovary with two feathery stigmas. All grasses are wind pollinated except the few that are self pollinated within closed florets (cleistogamous). The lemma has a midrib and a number of smaller "veins" or "nerves" running roughly parallel to it, but converging toward the tip. The midrib of the lemma may be prolonged into a beard or bristle, called an awn. Rarely the lateral nerves also protrude. If the lemma is prominently folded along the midrib, it is said to have a keel. The hardened lower end of the lemma is called a callus. The palea always has two veins near the sides, but lacks a midrib. In some grasses the palea is small or lacking. Usually the spikelet has a stalk or pedicel, or this may be absent, as in wheat and rye, and then the spikelet is said to be sessile. Usually spikelets break up at maturity into individual florets, each of which will then bear a segment of the rachilla. Some spikelets, like those of switch grass and foxtail grass, do not break up, but are shed from the plant whole.

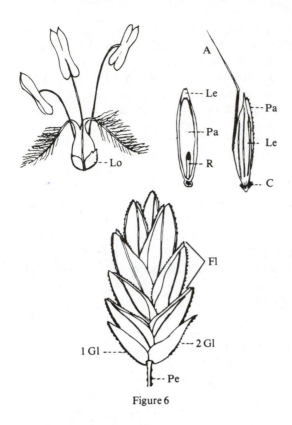

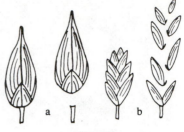

Figure 7

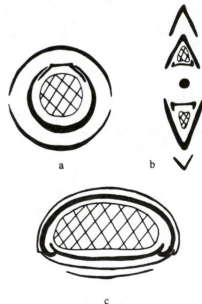

Figure 6

Often it is necessary to determine at what points the spikelets break or disarticulate. When the spikelets are mature and dry, they will disarticulate naturally, but if one has a rather immature plant, it may be necessary to tease the spikelets apart with needles and tweezers or with the finger nail in order to tell where the disarticulation will occur. There are two general types of disarticulation: below the glumes (Fig. 7, a), and above the glumes (Fig. 7, b). Spikelets which disarticulate below the glumes leave nothing on the plant except the stubs of the pedicels. Those which disarticulate above the glumes leave them on the plant. Spikelets of this type usually disarticulate between the florets as well.

Another feature of the spikelet which we may need to know is its shape in cross section. Spikelets may be round in cross section as in Fig. 8, a; flattened from the sides of the glumes and lemmas (laterally compressed), as in Fig. 8, b; or flattened from the backs of the glumes and lemmas (dorsally compressed), as in Fig. 8, c.

c

Figure 8

The kind of spikelet pictured in Fig. 6 is probably the basic type, from which reduced or more complicated sorts have been derived by various changes. In studying other types of grass spikelets, we should mentally compare them with the basic type in order to decide which parts have been modified or eliminated.

The following types of changes are common, and they characterize large groups of grasses:

1. The glumes may be large, covering the whole spikelets (oats and its relatives).

2. The upper florets may be eliminated, so that the spikelet is 1 flowered (redtop, timothy, and their relatives).

3. The lower florets may be sterile and much reduced in size, the upper one remaining fertile (canary grass, foxtail).

4. The glumes may be reduced to little ridges on the tip of the pedicel (rice, cut grass).

5. Either stamens or pistil may be eliminated, giving rise to spikelets or plants of one sex (salt grass, creeping lovegrass, Texas bluegrass, corn, sorghum, wild rice).

How to Collect and Study Grasses

Probably there is not a county in the United States where less than fifty to one hundred different species of grasses are to be found. Some will be very common and conspicuous, but others will be rare and hard to find. At first, all may look rather similar, so that sharp observation will be needed to detect even all of the common grasses around us. Don't be afraid to get down on hands and knees and crawl to get a good look. Each sort of habitat will have its own grasses: look in prairie, woodland, marsh, bog, ditches, corn or cotton fields, deserts, mountain meadows, or alpine summits, and you will probably be rewarded with a different set of species each time. Even in the same locality, new species come into flower throughout the spring, summer, and fall.

Grasses are easy to collect and prepare, so one should always take care to make good specimens, which will be a pleasure to study later. The tools needed for collecting grasses are simple. First, you will need some sort of digger, so that you can get the important underground parts of the plants. I use a long, stout screwdriver, but large hunting knives, geologist's picks, or entrenching tools are also satisfactory. Whatever tools you use, be sure to get the parts of the plant that lie below the ground level. Frequently an otherwise good specimen becomes very difficult to identify because the collector has neglected these structures.

After digging a specimen from the ground, one should knock the soil from the roots or wash them clean. If the plant is too bulky to press flat it may be subdivided and some of the extra inflorescences saved to provide spikelets for dissection. Each specimen is placed in a single folded newspaper sheet (12 x 16 inches as folded) for drying. Information which you would like to keep, such as the location, type of habitat, and the date, should be written on the margin of the paper. If the specimen is too long to fit in a folder, it may be doubled back one or more times. Crumpled or tangled parts should be smoothed out. The bent stems can be held in place by little slips of card with a slit cut in each (Fig. 9). The grass specimens in their paper folders may be carried between sheets of beaverboard or plywood, with a light strap around the bundle, for periods as long as a day, before it is necessary to dry them.

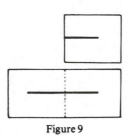

Figure 9

Final drying of the specimens is accomplished by placing them, in their folders, between 12 x 18 inch blotters made of builders' deadening felt (obtainable from lumber yards, in rolls), or between thick pads of newspapers. The specimens must be kept under pressure until dry, either by placing boards on the sides of the bundle and strapping it tightly, or by placing heavy weights on top of the bundle. Each day the damp blotters or newspaper pads must be removed and replaced with dry ones. The damp blotters may be dried by laying them out in the sun on dry paving (not grass) for a short while. In wet weather, the blotters can be dried cautiously in a warm oven. Usually grass specimens dry in a few days. After drying, they may be handled in the paper folders, but they will keep better if they are mounted on paper.

One may mount specimens in large scrapbooks, or better still, on standard herbarium sheets which may be purchased from biological supply houses. Specimens may be glued to paper by placing them momentarily on a large sheet of glass covered with thin glue, brushed to a thin uniform layer. Fish glue, carriage glue, or tin paste are satisfactory for this purpose. Each specimen, after gluing, is then dropped onto a sheet of paper. Since grasses are often quite waxy, they do not always stick well and should also be sewed to the sheet with string or fastened down with narrow strips of gummed cloth tape (Fig. 10). Do not use cellulose tape, since it becomes sticky and brittle with age or pulls loose. A label, bearing the name of the plant, the place and date of collection, and any other pertinent information, should be glued in the lower right corner of the sheet. Loose spikelets or other small parts may be placed in small coin envelopes glued to the sheet.

When considerable numbers of specimens are needed, for use in classes or for display, they may be preserved by tying them in sheaves and allowing them to hang head down until dry.

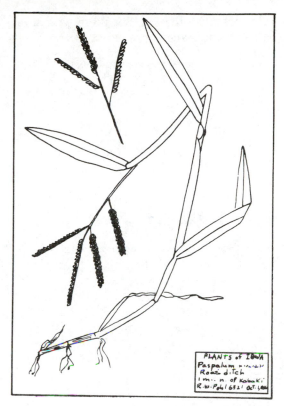

Figure 10

The equipment needed to study grasses is simple and mostly very inexpensive. However, thumbs are no substitute for dissecting tools. You will need at least two well-sharpened dissecting needles. These should be kept sharp by a fine-grained oilstone, such as a hard Arkansas Case or Behr-Manning stone. A micro-scalpel (Fig. 11) is needed for cutting small spikelet parts. This instrument can be made from a two-inch piece cut from the end of a jig-saw blade. One end of this, after sharpening, is thrust into a wooden dissecting needle handle. The other end can be shaped with a wire nipper and then sharpened to a cutting edge on the oilstone. Since fingers are bigger than florets, it is essential to have a good pair of tweezers. These should have angle-tips, which must meet perfectly. A razor blade and a metric ruler complete the kit. For those who always lose the ruler, one is printed at the end of the volume.

Figure 11

While it is possible to study grasses with a hand lens, a low power binocular microscope is a great help, since it frees both hands. Relatively low priced ones are now available. If you must use a hand lens, I suggest that you fasten it to an improvised stand.

Dry grass spikelets are often difficult to dissect, as they are stiff and tend to fly away when prodded. A wetting solution can be made from 25% wood alcohol, with the addition of several per cent of strong liquid detergent. Keep this in a dropper bottle and apply as needed. It will also soften leaves so they may be moved without breaking.

General References

The following short list includes some of the principal books useful for identification or study of American grasses. The list is not intended to be complete, and some books which are old or unavailable have been omitted. You will find that a book on the grasses of your own state, or an adjacent one, will often make identification easier than such comprehensive works as Hitchcock's manual, which covers a large and diverse area.

Beetle, A.A. 1971. Grasses of Wyoming. Wyo. Agric. Exp. Stn. Research Journal 39: 1-151. Keys, descriptions, illustrations, maps.

Blomquist, H.L. 1948. The grasses of North Carolina. Duke Univ. Press, Durham. vi plus 276 pp. Keys, descriptions and illustrations.

Booth, W.E. 1964. Agrostology. Montana State, Bozeman. 222 pp. Structure and classification of grasses; no keys or descriptions of individual species.

Chase, Agnes. 1959. First book of grasses. Smithsonian Institution Press, Washington, D.C. Good, well-illustrated explanations of spikelet structure for the various tribes.

Core, E.L., E.E. Berkley, and H.A. Davis. 1944. West Virginia grasses. Bull. 313, W. Va. Agric. Exp. Stn. 96 pp. Keys, descriptions, illustrations.

Deam, C.C. 1929. Grasses of Indiana. Pub. 82, Ind. Dep. Conservation, Indianapolis. Keys, descriptions, maps, illustrations.

Dore, W.G. and A.E. Roland. 1942. The grasses of Nova Scotia. Proc. N.S. Inst. Sci. XX: 177-288. Keys, discussions, illustrations, maps.

Fassett, N.C. 1951. Grasses of Wisconsin. Univ. of Wis. Press, Madison. 173 pp. Keys, maps, illustrations.

Featherly, H.I. 1946. Manual of the grasses of Oklahoma. Bull. 21, Okla. A & M College. 137 pp. Keys, descriptions, illustrations.

Gates, F.C. 1937. Grasses in Kansas. Rept. Kans. St. Board of Agric. Vol. LV, No. 220-A. 349 pp. Keys, descriptions, illustrations of species, maps.

Gould, F.W. 1965. Grasses of the Texas Coastal Bend. Texas A & M Univ. Press, College Station. 189 pp. Keys and illustrations.

Gould, F.W. 1968. Grass systematics. McGraw-Hill, New York. 382 pp. A modern textbook on agrostology.

Gould, F.W. 1975. The grasses of Texas. Texas A & M Univ. Press, College Station. 653 pp. Keys, descriptions, illustrations.

Harrington, H.D. 1946. Grasses of Colorado. Colorado A & M College, Ft. Collins. Mimeographed. 167 pp., plus index. Keys and descriptions, no illustrations.

Hitchcock, A.S. 1936. The genera of grasses of the United States, with special reference to the economic species. U.S. Dept. Agric. Bull. 772, revised ed. Supt. of Documents, Washington, D.C. 302 pp. Keys to genera, illustrations, discussions of important species.

Hitchcock, A.S. 1951. Manual of the grasses of the United States. U.S. Dep. Agric. Misc. Publ. 200, revised ed. (by Agnes Chase). Supt. of Documents, Washington, D.C. Abundantly illustrated. This is a very important publication on American grasses, but rather large and complex for the beginner.

Hubbard, Wm. A. 1955. The grasses of British Columbia. Handbook 9, B.C. Prov. Museum. 205 pp. Illustrated.

Kucera, Claire L. 1961. The grasses of Missouri. Univ. Missouri Press, 241 pp. Illustrated.

McClure, F.A. 1966. The bamboos: a fresh perspective. Harvard Univ. Press, Cambridge, Mass. 347 pp. General discussions of bamboo structure, reproduction, propagation and classification.

McClure, F.A. 1973. (*Ed.* T.R. Soderstrom). Genera of bamboos native to the New World. Smithsonian Contrib. to Botany 9: 1-347. Smithsonian Institution Press, Washington, D.C.

Mosher, Edna. 1918. The grasses of Illinois. Ill. Agric. Exp. Stn. Bull. 205: 261-425. Keys, descriptions, illustrations.

Norton, J.B.S. 1930. Maryland grasses. Md. Agric. Exp. Stn. Bull. 323. Keys, brief descriptions, key to vegetative characteristics of grasses.

Pohl, Richard W. 1966. The grasses of Iowa. Iowa grasses of Pennsylvania. Am. Midland Naturalist 38: 513-604. Keys and habitat notes, no descriptions or illustrations.

Pohl, Richard W. 1966. The grasses of Iowa. Iowa State J. of Sci. 40: 341-566. Keys and notes, illustrations, maps.

Pool, Raymond J. 1948. Marching with the grasses. xii plus 210 pp. Univ. of Nebraska Press, Lincoln. Economic botany of grasses, not useful for identification of individual genera or species.

Rotar, Peter P. 1968. Grasses of Hawaii. Univ. of Hawaii Press, Honolulu. Keys to tribes and genera, illustrations.

Silveus, W.A. 1933. Texas grasses. Published by author, San Antonio. 782 pp. Illustrations, keys, and descriptions.

U.S. Dep. Agric. 1948. Grass: The yearbook of agriculture 1948. Supt. of Documents, Washington, D.C. Numerous articles on grasses, legumes, grasslands. One section is on common agricultural grasses.

U.S. Forest Service. 1937. Range plant handbook. Supt. of Documents, Washington, D.C. xxvi plus 512 pp. One section on grasses. Illustrations, detailed notes on structure, uses by grazing animals, range, etc. No keys.

Recognizing Grass Tribes

The basic separation of the grass subfamilies and tribes is made upon microscopic characters, but many of these groups can be recognized by their more obvious features that require little or no magnification. It is helpful to know some of these, and they are summarized below.

Subfamily Bambusoideae

All of these grasses have woody perennial stems. The main stems do not have foliage leaves. The leaf blades, which are relatively small, are all found upon small lateral branches. The bamboos seldom bloom. The native and introduced species are all restricted to the southern half of the United States. Fig. 12.

Figure 12

Subfamily Oryzoideae

Mostly aquatic grasses with single florets and much reduced glumes.

Tribe Oryzeae. Rice tribe. Spikelets strongly laterally flattened; flower perfect. Fig. 13.

Figure 13

Tribe Zizanieae. Wild rice tribe. Aquatic grasses; spikelets tend to be round and are always unisexual. Fig. 14.

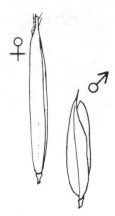

Figure 14

Figure 16

Subfamily Pooideae

This group includes many of the common temperate zone grasses. As now defined, it includes far fewer genera than given by Hitchcock. The more readily recognizable tribes are given below.

Tribe Poeae. Spikelets several-flowered, in panicles; glumes short; lemmas with 5 or more nerves. Bluegrasses, brome grasses, fescues, orchard grass, etc. Fig. 15.

Tribe Aveneae. Spikelets several-flowered, in panicles; glumes nearly or quite as long as the whole spikelet; awn, when present, from the back of the lemma or from a split tip. Oats, Junegrass. Fig. 17.

Figure 17

Figure 15

Tribe Agrostideae. Spikelets single-flowered, usually quite small, in a panicle; compression lateral. See also the tribe Sporoboleae, distinguished only on micro-characters. Fig. 18.

Tribe Triticeae. Spikelets borne on a balanced or symmetrical spike, 1 or more at each node; florets 1-several; leaf sheaths often bearing auricles. Wheat, rye, barley, quackgrass are examples. Fig. 16.

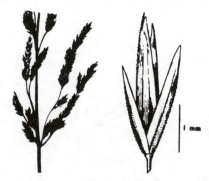

Figure 18

Tribe Stipeae. Spikelets single-flowered, round in cross section; glumes about as long as the body of the lemma; floret hard, cylindrical, with a hard, sharp callus and a stout, usually twisted awn; inflorescence a panicle. Fig. 19.

Figure 19

Tribe Phalarideae. Spikelets with long glumes, concealing the florets; florets 2-3, the lower 2 sterile or staminate, often very minute, all falling as a unit from the glumes; inflorescence a panicle. Fig. 20.

Figure 20

Subfamily Arundinoideae

Tribe Arundineae. Giant grasses, with stems 1-6 m tall and large, plumelike panicles, the spikelets covered with silky hairs; florets several; spikelets laterally compressed. Fig. 21.

Figure 21

Tribe Aristideae. This tribe contains only the needle grasses of the genus *Aristida*. Spikelets single-flowered, the floret hard and cylindrical, with a sharp callus and 3 awns. Fig. 22.

Figure 22

Subfamily Chloridoideae

These grasses were separated from the large subfamily Festucoideae of Hitchcock on microscopic characters. They are mostly steppe, desert, and warm climate grasses. Those which resemble members of the tribe Poeae (Festuceae) can usually be recognized by having 3 or fewer nerves on the lemma.

Tribe Eragrosteae. Inflorescence a panicle or a group of 1-sided spikes; spikelets with several to many fertile florets; lemmas 3 nerved. Fig. 23.

Figure 23

Tribe Sporoboleae. Inflorescence a panicle; spikelets with a single floret; lemma with 1-3 nerves. Similar to the Agrostideae, but differing in micro-characters. Fig. 24.

A

Figure 24

Tribe Chlorideae. Inflorescence of 1—many 1-sided spikes; spikelets with 1 fertile floret and 1 or more rudimentary ones above or below it. Fig. 25.

Figure 25

Tribe Pappophoreae. Desert grasses; inflorescence a panicle; lemmas many nerved, the apex cut into many lobes or many-awned. Fig. 26.

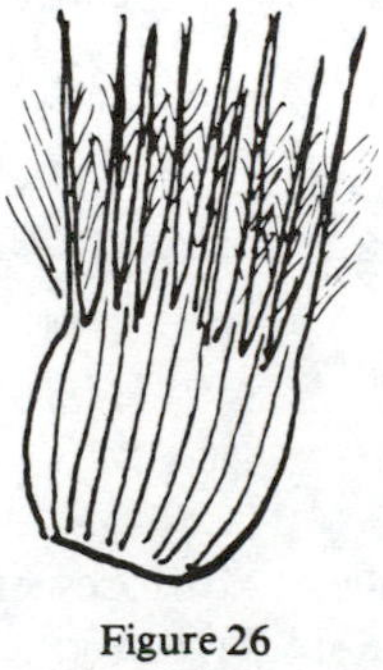

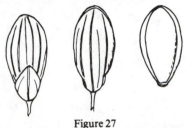

Figure 26

Subfamily Panicoideae

Spikelets dorsally compressed, with 1 fertile floret, a sterile or staminate one and a pair of glumes below it.

Tribe Paniceae. First glume short or lacking; second glume as long as the spikelet; glumes and sterile lemma thin; fertile floret leathery or rigid; inflorescence a panicle or of 1-sided racemes or spikes. Fig. 27.

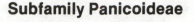

Figure 27

Tribe Andropogoneae. Both glumes as long as the spikelet and concealing the inner parts; sterile and fertile lemmas thin and delicate; spikelets usually disposed in pairs at each rachis joint; both may be stalked or one may be sessile. The stalked spikelet tends to be reduced or absent. Many variations on the pattern of arrangement occur. Fig. 28.

Figure 28

How to Use the Keys

The keys in this work are of the usual dichotomous "bracket" type that is used in other books of this series. At each point in the key, the user must consider the two choices offered and pick the better one. The guide numbers at the right hand margin will eventually lead the user to a genus, and a further key under the genus goes on to a species. For greater convenience, the genera are first divided into six major groups (page 18), and a separate key is provided for each of these. It is very important to make the primary separation very carefully. You should read and understand each entire key lead. Especially with small-flowered plants like grasses, it will not do to take a brief look at the first phrase of a key lead and then make a snap decision. Note particularly the little words "and" and "or." They make a great deal of difference in meaning. At least in the beginning, it will be necessary to use the glossary or the introduction to be sure of the meaning of terms. Once you are familiar with the language used to describe grass structures, it will not be any more troublesome than the ordinary terms used to describe other flowering plants.

A summary of the more conspicuous and easily recognized grass tribes is presented on page 13 and the following pages. With some practice, you will be able to recognize most of these tribes on sight, and this should be a great aid in deciding whether you are close to your goal in performing an identification.

Finally, if you have never identified grasses before, you are venturing into a foreign land, and familiar landmarks will be few. When you have learned a few of the most conspicuous kinds of grasses, they will serve as guides. If you are familiar with some of the common lawn, crop, and weedy grasses, it may be helpful to look them up in the index and look at the descriptions and pictures. If your progress for the first few days is slow, be of good cheer. We have all traveled the same road.

Preliminary Key to Separate Major Groups of Grasses

1a Stems woody and perennial, several to many m tall; main culms bearing only bladeless sheaths which soon fall; leaves with blades borne only on smaller branches from the nodes of the culms; plants flowering only at long, irregular intervals of years (Bamboos: Subfamily BAMBUSOIDEAE). Fig. 29 . . . . 1. (p. 44). *Arundinaria*

Figure 29

1b Stems not woody, surviving only one growing season, bearing ordinary foliage leaves, flowering annually 2

2a Spikelets not themselves spiny nor enclosed in spiny adhering structures . . . 3

2b Spikelets bearing straight or hooked spines on the glumes, or enclosed in spiny structures. Fig. 30 . . . . (p. 20) KEY I

Figure 30

3a Giant grasses, the culms thick, 2 or more m tall, corn-like or reed-like, sometimes with large plume-like panicles . . . . (p. 20) KEY II

3b Grasses of small to moderate stature, rarely more than 1-2 m tall. 4

4a Spikelets (at least the pistillate ones in unisexual forms) enclosed or partly hidden in hardened rachis internodes or bead-like structures, or at times nearly completely hidden among foliage leaves. Fig. 31 . . . . (p. 22) KEY III

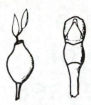

Figure 31

4b Spikelets not enclosed or hidden, usually visible in inflorescences raised above the foliage. 5

5a Spikelets disarticulating above the glumes, which are left on the plant after maturity. Fig. 32, B 6

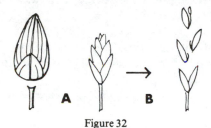

Figure 32

5b Spikelets either disarticulating below the glumes and falling as units, or dropping from the plants as groups or clusters, or remaining on the plants, only the glumes and lemmas falling. Fig. 32, A. . . . .
. . . . (p. 35) KEY VI

6a Spikelets consisting of only 2 glumes and a single fertile floret; rachilla may sometimes be prolonged behind the palea, but there is no evidence of a second floret at its apex . . . .
. . . . (p. 23) KEY IV

6b Spikelets with 2 or more florets, some of which may be sterile or rudimentary and borne above or rarely below a fertile floret. . . . .
. . . . (p. 26) KEY V

General Pictured Keys to the Genera of American Grasses

KEY I

Spikelet Groups Covered with Straight or Hooked Spines, and Falling as Units from the Rachis of the Spike

1a Spikelet groups enclosed in a spiny bur made up of flat, stiff plates; spines straight, minutely barbed............ .... 123. (p. 177) *Cenchrus*

1b Spikelet groups not enclosed, the lower glumes of the lower spikelets covered with hooked spines .... 108. (p. 153) *Tragus*

KEY II

Giant Grasses, the Culms 2 M or More Tall; Plants of Corn-Like or Reed-Like Aspect, Often with Large, Silky Panicles

1a Inflorescence dense, cylindrical, cattail- or foxtail-like, with the spikelets inter- mixed with numerous sterile bristles.... 2

1b Inflorescence not dense and cylindrical, lacking sterile bristles.............. 3

2a Bristles about as long as the spikelets, and dropping attached to them; cultivated crop .... 122. (p. 176) *Pennisetum*

2b Bristles several times as long as the spikelets; fertile lemma hard, shiny, fall- ing from the spikelet, the glumes and bristles remaining on the plant......... .... 121. (p. 173) *Setaria*

3a Plants with separate staminate and pistillate inflorescences, the staminate in terminal panicles, the pistillate axillary, hidden in sheaths, only the styles (silks) protruding; corn-like plants.......... .... 146. (p. 192) *Zea*

3b Plants either perfect-flowered, or with flowers of the 2 sexes in the same in- florescence 4

4a Disarticulation below the glumes, entire

spikelets or spikelet groups falling from the inflorescence when mature 5

4b Disarticulation above the glumes, which remain on the plant 14

5a Spikelets falling in pairs, with attached rachis internodes 9

5b Spikelets falling from the inflorescence singly, without attached rachis internodes 6

6a Spikelets lacking visible glumes, consisting of a single staminate or pistillate floret; all aquatic plants. 7

6b Spikelets with glumes, perfect-flowered; plants of dry or wet habitats 8

7a Pistillate spikelets long-awned, all in an erect cluster at the tip of the inflorescence; staminate spikelets awnless, drooping, borne on spreading lower branches . . . . 4. (p. 46) *Zizania*

7b Pistillate and staminate spikelets intermixed on the same inflorescence branches, both short-awned or awnless . . . . 5. (p. 46) *Zizaniopsis*

8a Spikelets strongly laterally compressed, sessile, borne in 2 rows in dense, 1-sided spikes; inflorescence not silky-hairy. Fig. 33. . . . . 92. (p. 140) *Spartina*

Figure 33

8b Spikelets dorsally compressed, pedicellate in elongated racemes; inflorescence a silky-hairy panicle . . . . 130. (p. 182) *Miscanthus*

9a Inflorescence a panicle or a cluster of racemes or rames; spikelets either perfect-flowered or sterile, not all unisexual 10

9b Inflorescence composed of 1-several stiff erect spikes, each with a series of bony internodes at its base, each internode enclosing a single pistillate spikelet; upper part of the rachis flattened, each node bearing 2 staminate spikelets . . . . 145. (p. 192) *Tripsacum*

10a Inflorescence branches each bearing numerous spikelet pairs 12

10b Inflorescence branches bearing short groups (rames), each consisting of 1-3 (-5) pairs of spikelets; many such groups forming a compound panicle 11

11a Each spikelet pair consisting of a sessile, awned, perfect-flowered spikelet and a

pedicellate, awnless staminate spikelet; crop plants or weeds . . . .
. . . . 137. (p. 187) *Sorghum*

11b   Each pair consisting of a sessile, awned, perfect-flowered spikelet, accompanied by a hairy sterile pedicel, the spikelet rudimentary or usually absent . . . .
. . . . 138. (p. 188) *Sorghastrum*

12a   Spikelets awnless; cultivated crop . . . . . .
. . . . 131. (p. 182) *Saccharum*

12b   Spikelets with awns; wild plants 13

13a   Spikelets of the pair equal, both awned, fertile . . . .
. . . . 129. (p. 181) *Erianthus*

13b   Spikelets of the pair unlike, the sessile one perfect-flowered, awned, the pedicellate spikelet staminate or sterile, awnless . . . .
. . . . 133. (p. 184) *Andropogon*

14a   Lemmas glabrous, the rachilla internodes densely silky; wild plants of wet ground and marshes . . . .
. . . . 67. (p. 111) *Phragmites*

14b   Lemmas hairy; cultivated plants, usually on dry ground 15

15a   Leaves mostly basal, long and narrow, drooping; plants forming large, fountain-like clumps . . . .
. . . . 68. (p. 112) *Cortaderia*

15b   Leaves equally spaced along the culm, broad and with clasping blade bases . . . .
. . . . 66. (p. 111) *Arundo*

KEY III

Spikelets Concealed in Hard Bony Structures, or Completely Hidden in Leaf Sheaths, or Borne Among Foliage Leaves and Scarcely Distinguishable from Them

1a   Low grasses, 5-10 cm tall, forming mound-like patches, or creeping by stolons. 2

1b   Erect, tall grasses, not stoloniferous, mostly more than 1 m tall 3

2a   Plants stoloniferous; leaf blades soft; spikelets borne in hardened, green-crowned basal structures. Fig. 34. . . . . . .
. . . . 100. (p. 148) *Buchloë*

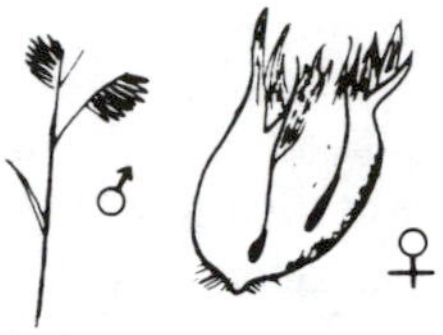

Figure 34

2b   Plants forming circular mounds, not stoloniferous; leaf blades stiff, sharp-pointed, recurved; spikelets hidden among upper leaves. . . . .
. . . . 101. (p. 149) *Munroa*

3a   Staminate spikelets borne in a terminal panicle (tassel); pistillate spikelets borne on axillary spikes covered with leaf sheaths, only the styles (silks) protruding.
. . . . 146. (p. 192) *Zea*

3b   Staminate and pistillate spikelets borne in the same inflorescence, none concealed in leaf axils 4

4a Plants producing a single spherical hard bead at the tip of each peduncle; styles of the pistillate flower and the stalk of the staminate spikelet cluster protruding from the mouth of the bead; cultivated ornamental. . . . . **144. (p. 191)** *Coix*

4b Peduncles bearing 1 to several stiff erect spikes; basal several internodes of each spike hard, cylindrical, each concealing a single pistillate spikelet; terminal portion of each spike with a flattened rachis, each node bearing a pair of staminate spikelets; wild plants. Fig. 35 . . . . **145. (p. 192)** *Tripsacum*

Figure 35

KEY IV

Spikelets Disarticulating Above the Glumes, Which Remain on the Plant; 1 Floret, No Sterile or Rudimentary Florets

1a Inflorescence a panicle, either open or dense and cylindrical, the spikelets never arranged in definite rows 2

1b Inflorescence a single balanced spike, or a group of 1-sided spikes or spikelike racemes; spikelets always in uniform rows 22

2a Glumes reduced to a minute cupule at the apex of the pedicel; spikelets strongly laterally compressed. Fig. 36 3

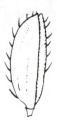

Figure 36

2b Glumes evident, either short or long; spikelets either laterally compressed, dorsally compressed, or round in cross section. 5

3a Spikelets with 2 sterile lemmas at the base of the fertile floret, these about half as long as the floret; cultivated crop; aquatic. Fig. 37 . . . . **2. (p. 45)** *Oryza*

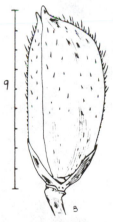

Figure 37

3b Spikelets lacking sterile lemmas at the base of the fertile floret 4

4a Lemmas 3 nerved, awned . . . .
. . . . 86. (p. 131) *Muhlenbergia*

4b Lemmas with more than 3 nerves, awnless. Fig. 36 . . . .
. . . . 3. (p. 45) *Leersia*

5a Lemma with 3 awns, the 2 lateral ones sometimes shorter than the central one . .
. . . . 72. (p. 114) *Aristida*

5b Lemma awnless or with 1 awn 6

6a Floret hard, round in cross section, usually with a sharp, bearded callus and a stiff, often twisted awn. Fig. 38 7

Figure 38

6b Floret soft or leathery, laterally or dorsally compressed, awnless or with a delicate awn. Fig. 39 9

Figure 39

7a Edges of lemma turned inward, fitting into the groove between the prominent, ridge-like keels of the palea . . . .
. . . . 64. (p. 110) *Piptochaetium*

7b Edges of the lemma overlapping, not inturned; palea flat, not grooved 8

8a Awn firmly attached to the lemma, its basal segment spirally twisted; callus usually sharp . . . .
. . . . 62. (p. 106) *Stipa*

8b Awn readily separating from the lemma, not spirally twisted; floret plump, its callus usually blunt . . . .
. . . . 63. (p. 108) *Oryzopsis*

9a Fertile floret bearing at its base 2 small appressed scales (sterile lemmas) that fall attached to the smooth, shiny, hard lemma. Fig. 40 . . . .
. . . . 35. (p. 81) *Phalaris*

Figure 40

9b Fertile floret without 2 basal scales. . . . 10

10a Rachilla prolonged behind the palea as a little bristle (CARE: This is sometimes hidden among hairs). Fig. 41 11

Figure 41

10b Rachilla not evident behind the palea . . 15

11a Glumes less than one-tenth as long as the long-awned floret . . . . 61. (p. 106) *Brachyelytrum*

11b Glumes more than one-fourth as long as the awned or awnless floret 12

12a Glumes swollen above the base; awn several times as long as the minute floret; small tufted annual . . . . 39. (p. 86) *Gastridium*

12b Glumes not swollen above the base; awn absent or scarcely longer than the lemma; perennials 13

13a Spikelets 10-20 mm long; coarse, tough grasses of coastal dunes or along the Great Lakes. . . . . 36. (p. 81) *Ammophila*

13b Spikelets less than 10 mm long; grasses of various habitats 14

14a Lemmas blunt, awnless, with 3 parallel nerves . . . . 19. (p. 73) *Catabrosa*

14b Lemmas acute, awned, 5 nerved, nerves converging . . . . 37. (p. 82) *Calamagrostis*

15a Floret dorsally compressed, smooth, shiny, rigid, awnless, concealed between the equal long glumes . . . . 46. (p. 89) *Milium*

15b Floret not dorsally compressed, awned or awnless, not hard and shiny . . . . 16.

16a Callus of the floret bearing an abundant

tuft of long, erect hairs, about half as long as the awnless lemma; coarse, rhizomatous sand-binding grasses . . . . 91. (p. 140) *Calamovilfa*

16b Callus of floret glabrous or with a few minute hairs; grasses of various habitats 17

17a Spikelets U-shaped, very flat, the equal, strongly keeled glumes square at the apex, only the midribs protruding as short awns; panicle dense, pencil-shaped; hay and meadow grass . . . . 45. (p. 88) *Phleum*

17b Spikelets not U-shaped; glumes tapering to the tip; inflorescences of various shapes 18

18a Glumes equal, both longer than the floret 24

18b Glumes unequal or equal, one or both shorter than the floret 19

19a Nerves of the awnless lemma all 3 silky; palea hairy between the keels . . . . 90. (p. 139) *Blepharoneuron*

19b Nerves of the awned or awnless lemma not hairy; palea not hairy. 20

20a Panicle short, thick, dense, partly hidden at the base in the upper leaf sheath; weed. . . . . 89. (p. 139) *Heleochloa*

20b Panicle not dense and hidden, either exposed on a peduncle, or very slender and concealed in the leaf sheath 21

21a Lemmas plainly 3 nerved, usually awned

from the tip, mostly with a tuft of short hairs on the callus . . . .
. . . . 86. (p. 131) *Muhlenbergia*

21b   Lemmas 1 nerved, rarely with faint lateral nerves, awnless, without callus hairs . . . .
. . . . 88. (p. 136) *Sporobolus*

22a   Inflorescence a single balanced spike on each peduncle; lemma awned. . . . .
. . . . 41. (p. 86) *Scribneria*

22b   Inflorescence of several to many 1-sided spikes; lemma awnless 23

23a   Spikes borne in a raceme along a slender rachis; tufted plants of the Great Plains; native . . . .
. . . . 97. (p. 145) *Schedonnardus*

23b   Spikes borne in a whorl at the tip of the peduncle; plants producing stolons and rhizomes; lawns and pastures, southern U.S. . . . .
. . . . 98. (p. 145) *Cynodon*

24a   Glumes tapering to slender awn points; lemma with 3 evident nerves; rhizomes densely covered with overlapping scales. .
. . . . 86. (p. 131) *Muhlenbergia*

24b   Glumes not awned, the tip acute; lemma faintly 5 nerved; rhizomes, if present, not densely scaly . . . .
. . . . 38. (p. 83) *Agrostis*

KEY V

Spikelets Disarticulating Above the Glumes; Florets 2 or More, Some of

Which May Be Sterile or Modified And Borne Above or Below Any Fertile Ones. Fig. 42

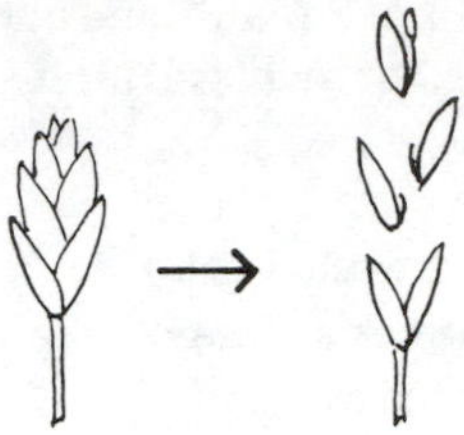

Figure 42

1a   Spikelets hidden among the stiff recurved leaf blades and similar to them . . . .
. . . . 101 (p. 149) *Munroa*

1b   Spikelets borne in an inflorescence above the leaves. 2

2a   Spikelets borne in rows in a single spike or raceme, or in 1 to many 1-sided spikes or racemes 3

2b   Spikelets borne in an open or dense panicle, each spikelet on a pedicel, not borne in definite rows 20

3a   Spikelets few, large, drooping, awned, pedicellate in a raceme; sheaths with united edges . . . .
. . . . 60. (p. 105) *Pleuropogon*

3b   Spikelets numerous, erect or spreading, sessile or nearly so; sheaths with overlapping edges 4

4a   Spikelets unisexual, the two sexes very different, the pistillate ones many-flowered, each floret bearing 3 long awns; staminate spikelets awnless . . . .
. . . . 85. (p. 131) *Scleropogon*

4b Spikelets perfect-flowered, all alike 5

5a Spikes solitary, erect, bearing spikelets on opposite sides of the rachis (balanced spike). Fig. 43 6

Figure 43

5b Spikes 1 to many, 1-sided, variously arranged 11

6a Spikelets 1 at each node of the rachis . . . 7

6b Spikelets 2 or more at each node of the rachis. 10

7a Spikelets placed with one edge against the rachis, only one glume developed. Fig. 43. . . . . 11. (p. 61) *Lolium*

7b Spikelets placed with one flat side against the rachis; both glumes equally developed 8

8a Spikelets 2 flowered, the keels of the lemmas bearing a row of short, comb-like bristles . . . . 49. (p. 91) *Secale*

8b Spikelets with 3 or more florets; keels

of the lemmas not bearing a row of bristles. 9

9a Glumes tapering to a narrow tip; grain remaining in the floret at maturity; wild or cultivated forage grasses or weeds . . . . 47. (p. 89) *Agropyron*

9b Glumes blunt; grain falling free from the floret at maturity; cultivated crop . . . . 48. (p. 91) *Triticum*

10a Glumes usually reduced to short stubs or bristles; spikelets spreading perpendicular to the rachis; inflorescence a brush-like open spike. . . . . 52. (p. 95) *Hystrix*

10b Glumes well developed; spikelets erect, pressed against the rachis of a dense spike . . . . 51. (p. 92) *Elymus*

11a Inflorescence consisting of a single, strongly curved spike on each peduncle; southeastern U.S. . . . . 96. (p. 145) *Ctenium*

11b Inflorescence of 2 to many spikes on each peduncle 12

12a Spikelets with only a single perfect-flowered floret, with 1 or more modified sterile florets above it 13

12b Spikelets with 2 or more perfect-flowered or staminate florets 17

13a Fertile lemma bearing 3 long awns, at least twice as long as the body of the lemma. . . . . 93. (p. 142) *Chloris crinita*

13b Fertile lemma awnless or with a single long awn or 3 short ones 14

14a Spikes borne singly at each node of an elongated central rachis 15

14b Spikes borne in 1 to several whorls, with 2 to many spikes in each whorl 16

15a Spikes very long and slender, 10-20 cm long, the widely spaced spikelets parallel to the rachis; spikes numerous, forming an open, dome-shaped panicle. 95. (p. 144) *Gymnopogon*

15b Spikes thicker, less than 5 cm long, the closely packed spikelets placed at an angle to the rachis; spikes 1 to many . . . . 99. (p. 146) *Bouteloua*

16a Lemmas dark brown, nearly awnless; glumes blunt, bifid at the apex; spikelets placed perpendicular to the rachis . . . . 94. (p. 144) *Eustachys*

16b Lemmas light colored, awned; glumes tapering to the tips; spikelets appressed to the rachis at an acute angle . . . . 93. (p. 142) *Chloris*

17a Florets 2, staminate; low, creeping native perennial of the Great Plains . . . . 100. (p. 148) *Buchloë*

17b Florets more than 2, perfect-flowered 18

18a Inflorescence of 1-2 whorls of short thick spikes; lemmas strongly keeled, awnless 19

18b Inflorescence a panicle of numerous spikelike racemes, arranged in a raceme along a central rachis; spikelets minutely pedicellate, not strongly keeled . . . . 77. (p. 125) *Leptochloa*

19a Second glume with a short curved awn. . . . . 79. (p. 127) *Dactyloctenium*

19b Second glume awnless . . . . 78. (p. 126) *Eleusine*

20a Most of the branches of the dense, cylindrical panicle converted into sterile bristles; fertile floret hard, shiny, dorsally compressed, awnless Fig. 44 . . . . 121. (p. 173) *Setaria*

Figure 44

20b Panicle branches all bearing spikelets; fertile floret awned or awnless, not dorsally compressed 21

21a Lemma divided at its summit into 9 or more awns or lobes. Fig. 45 22

Figure 45

21b Lemmas with 0-3 awns. 24

22a Florets disarticulating separately; glumes much shorter than the spikelet. . . . .
. . . . 103. (p. 150) *Cottea*

22b Florets falling from the glumes as a group, not separating; glumes nearly as long as the spikelet. 23

23a Florets 3; awns fringed with hairs . . . . . .
. . . . 105. (p. 151) *Enneapogon*

23b Florets 4 or more; awns not hairy . . . . . .
. . . . 104. (p. 151) *Pappophorum*

24a Leaf sheaths with edges united, at least below the middle. Fig. 46, A 25

24b Leaf sheaths with overlapping, separate edges. Fig. 46, B 29

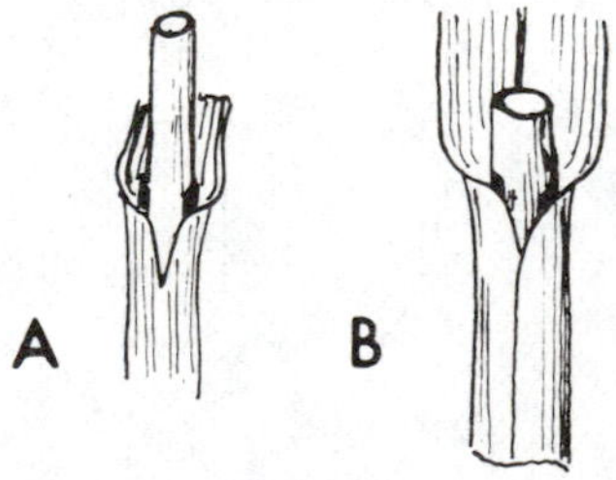

Figure 46

25a Lemmas awnless, blunt, with 5-9 prominent parallel nerves, not coming together at the tip; spikelets shattering very readily; plants of wet soil or shallow water. Fig. 47 . . . .
. . . . 59. (p. 102) *Glyceria*

Figure 47

25b Lemmas acute, awned or with split tips; nerves of the lemmas converging toward the tips; not aquatics 26

26a Panicle branches few, stiff; spikelets densely bunched near their tips, almost sessile; sheaths keeled; cultivated pasture grass . . . .
. . . . 20. (p. 73) *Dactylis*

26b Panicle branches not few and rigid; spikelets separate, on pedicels; wild or cultivated grasses. 27

27a Lemmas awned, with a conspicuous tuft of straight erect hairs on the callus. . . . . .
. . . . 58. (p. 102) *Schizachne*

27b Lemmas awned or awnless, without tufts of callus hairs 28

28a Upper several florets of the spikelet sterile, wrapped around each other, form-

ing a club-shaped structure Fig. 48......
................57. (p. 97) *Melica*

Figure 48

28b   Upper florets sometimes reduced in size, but not wrapped around each other......
...............7. (p. 47) *Bromus*

29a   Tall grasses, 1-5 m tall, with large plume-like panicles, the spikelets silky because of long hairs on the lemmas or rachilla 30

29b   Grasses of moderate size, usually 1 m or less tall; panicles not silky 32

30a   Lemmas hairy 31

30b   Lemmas glabrous, the rachilla internodes densely silky; plants of wet ground and marshes.........................
.............67. (p. 111) *Phragmites*

31a   Plants forming large clumps of long, narrow, drooping basal leaves; culms nearly leafless ....
.............68. (p. 112) *Cortaderia*

31b   Plants with leaves spaced equally along the culms; blades broad, with clasping bases.........................
...............66. (p. 111) *Arundo*

32a   Spikelets of 2 kinds, different in appearance, in the same or different inflorescences....................33

32b   Spikelets all similar (rarely unisexual) . 34

33a   Spikelets unisexual, borne in different inflorescences, the pistillate ones having several florets, each lemma with 3 long awns, the staminate ones awnless......
............85. (p. 131) *Scleropogon*

33b   Spikelets paired in the same inflorescence, one of each pair with 2-3 perfect-flowered florets, the rachilla disarticulating; the other with a non-disarticulating rachilla bearing many sterile, empty lemmas ....
...............21. (p. 73) *Cynosurus*

34a   All the florets dropping from the glumes as a unit.......................35

34b   Florets disarticulating from each other and dropping separately...........38

35a   Lower 2 florets reduced to little awnless scales, closely appressed to the base of the much larger, hard, shiny terminal floret. Fig. 49.........................
...............35. (p. 80) *Phalaris*

Figure 49

35b   Lower florets well developed, as large

as or larger than the terminal floret. Fig. 50 36

Figure 50

36a Florets 4, only the third fertile, the fourth reduced to an awned rudiment; desert grasses. . . . .
. . . . 83. (p. 130) *Blepharidachne*

36b Florets 3, the terminal one fertile, awnless, smaller than the lower 2, which are awned or awnless, staminate or sterile; grasses of moist areas, with the odor of vanilla 37

37a Lower 2 florets awned, sterile; glumes very unequal . . . .
. . . . 33. (p. 79) *Anthoxanthum*

37b Lower 2 florets awnless, staminate; glumes equal . . . .
. . . . 34. (p. 80) *Hierochloë*

38a Lemmas with 3 prominent nerves. 39

38b Lemmas with 5 or more nerves, which are usually not conspicuous. 47

39a Glumes at least three-fourths as long as the entire spikelet. 40

39b Glumes much shorter than the spikelet, usually reaching only to the tip of the lowermost floret 41

40a Tall grasses, the culms with several nodes and leaves; inflorescence an elongated, spikelike panicle . . . .
. . . . 80. (p. 127) *Tridens*

40b Low, tufted or sprawling desert grasses; culms with a single node . . . .
. . . . 81. (p. 128) *Erioneuron*

41a Lemmas with 1-3 awns or short awn-tips, or awnless but with a split tip 42

41b Lemmas awnless, the tip not split. 44

42a Lemmas lobed, the awn, if present, arising between the lobes 43

42b Lemmas not lobed, the short awn arising at the tip. Fig. 51 . . . .
. . . . 81. (p. 128) *Erioneuron*

Figure 51

43a Palea long-hairy on its upper half; inflorescence mostly hidden in the uppermost sheath. . . . .
. . . . 82. (p. 129) *Triplasis*

43b Palea not hairy; inflorescence exposed. . .
. . . . 80. (p. 127) *Tridens*

44a Lemmas with a tuft of hairs on the callus; rhizomatous sand-binding grass of the Great Plains . . . .
. . . . 84. (p. 130) *Redfieldia*

44b Lemmas lacking a tuft of callus hairs; habitats various, but not sand-binders **45**

45a Grain large, bottle-shaped, 5-6 mm long, forcing the stiff, strongly-keeled lemma and the palea apart at maturity and protruding between them. Fig. 52 . . . . **73. (p. 116)** *Diarrhena*

Figure 52

45b Grain not bottle-shaped nor forcing the floret open, not over 2 mm long; lemmas thin, soft **46**

46a Spikelets with 3 or more florets; lemmas pointed, the nerves converging; plants of dry or moist soil . . . . **76. (p. 117)** *Eragrostis*

46b Spikelets with 2 florets; lemmas with blunt tips and parallel nerves; aquatic grass, mostly in the northern Rocky Mountains. . . . . **19. (p. 73)** *Catabrosa*

47a Lemmas awned or with a split tip **48**

47b Lemmas awnless, never split at the tip **56**

48a Lemmas tapering to an awn at the tip, never split at the tip; glumes much shorter than the spikelet **69**

48b Lemmas awned from the back or from the notch between 2 teeth; glumes nearly as long as the spikelet **49**

49a Glumes and lemmas all 5-9 nerved; lemmas with a minute straight awn from a notched tip; diminutive annual desert grasses of the Southwest . . . . **70. (p. 113)** *Schismus*

49b Lemmas 5 nerved; awn well developed, twisted, arising from the back of the lemma or between 2 teeth; grasses of various climates **50**

50a Awn attached between 2 apical teeth . . . . **69. (p. 112)** *Danthonia*

50b Awn attached on the back of the lemma below the tip **51**

51a Spikelets drooping, over 2 cm long; glumes 7-9 nerved; annual crop or weeds . . . . **23. (p. 74)** *Avena*

51b Spikelets not drooping, less than 2 cm long; glumes with 5 or fewer nerves; wild or cultivated plants **52**

52a Glumes 3-5 nerved; florets 2-6; rachilla hairy; spikelets over 1 cm long . . . . **24. (p. 75)** *Helictotrichon*

52b Glumes 1-3 nerved; florets 2; rachilla not hairy; spikelets less than 1 cm long **53**

53a Lower floret staminate, with a bent,

twisted awn; upper floret perfect-flowered, with a short straight awn . . . . 25. (p. 75) *Arrhenatherum*

53b Both florets alike and perfect-flowered 54

54a Awn attached near the base of the lemma 55

54b Awn attached above the middle of the lemma. . . . . 28. (p. 77) *Trisetum*

55a Rachilla prolonged beyond the base of the upper floret as a minute hairy bristle . . . . 26. (p. 76) *Deschampsia*

55b Rachilla not prolonged beyond the second floret. . . . . 27. (p. 76) *Aira*

56a Plants dioecious, the spikelets of both sexes similar but borne on separate plants 57

56b Plants with perfect flowers. 59

57a Lemmas with numerous faint nerves; plants of salty or alkaline soil, on seacoasts or dry interior areas . . . . 75. (p. 117) *Distichlis*

57b Lemmas 5 nerved; plants ordinarily not on salty soil 58

58a Plants clump-forming, rarely with rhizomes; lemmas glabrous; plants of the Rocky Mountains . . . . 10. (p. 60) *Leucopoa*

58b Plants turf-forming by rhizomes; lemmas hairy on the keel or callus; Pacific coastal dunes or plains of the south central and southeastern states . . . . 17. (p. 64) *Poa*

59a Glumes nearly as long as the spikelet . . 60

59b Glumes much shorter than the spikelet, usually reaching only to the tip of the lowermost lemma 62

60a Lemmas with a conspicuous tuft of erect hairs on the callus, 7 nerved, with a ragged tip; rhizomatous marsh plants . . . . 14. (p. 63) *Scolochloa*

60b Lemmas without conspicuous tufts of hairs on the callus, tapering to a narrow tip; florets usually 2; plants tufted, not occurring in marshes 61

61a Rachilla internodes about one-third as long as the lemmas, long-hairy; florets 2 . . . . 28. (p. 77) *Trisetum*

61b Rachilla internodes very short, not long-hairy; florets 2-4 . . . . 30. (p. 78) *Koeleria*

62a Spikelets large, very flat, 1-1.5 cm wide, the lemmas strongly keeled, with many faint nerves; 1 or more of the lower florets empty; stamens 1 per flower . . . . 71. (p. 113) *Chasmanthium*

62b Spikelets not strongly flattened, less than 1 cm wide; lemmas not strongly keeled; lower floret fertile; stamens 3. 63

63a Lemmas nearly circular in outline,

awnless, very faintly nerved, oriented at right angles to the rachilla . . . .
. . . . 18. (p. 72) *Briza*

63b Lemmas longer than wide, appressed to the rachilla at an acute angle 64

64a Lemmas bearing a tuft of crinkled, cottony hairs on the callus. Fig. 53 . . . . . .
. . . . 17. (p. 64) *Poa*

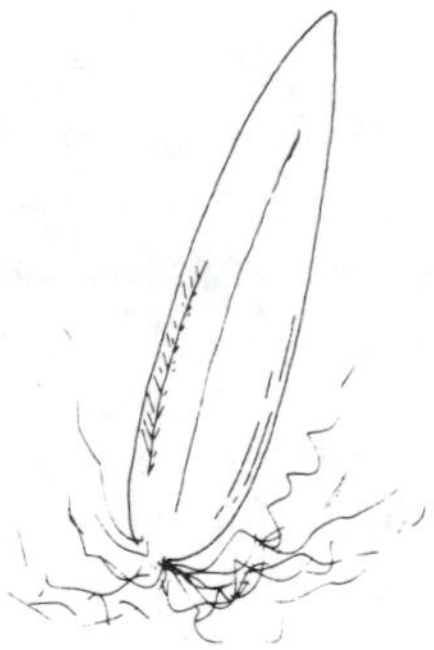

Figure 53

64b Lemmas without cottony hairs on the callus 65

65a Lemmas tapering to an acute tip or awn-tipped; nerves usually inconspicuous, usually converging toward the tip of the lemmas 66

65b Lemmas with blunt tips; nerves faint or conspicuous, running parallel toward the tip but not converging 68

66a Low annuals with small panicles, less than 10 cm long, the branches rigid and nearly simple; rachis and branches triangular; pedicels shorter than the 5-10 flowered spikelets; pedicels thick, rigid, triangular in cross section . . . .
. . . . 15. (p. 63) *Catapodium*

66b Annuals or perennials of various sizes; panicles small or large; rachis and branches usually round; pedicels thin and flexible, usually longer than the spikelets 67

67a Lemmas sharp-pointed or with a short awn-tip; leaf blades tapering to fine points; lemmas glabrous . . . .
. . . . 9. (p. 57) *Festuca*

67b Lemmas not sharp pointed; leaf blades with blunt, folded boat-shaped tips. Fig. 54 . . . .
. . . . 17. (p. 64) *Poa*

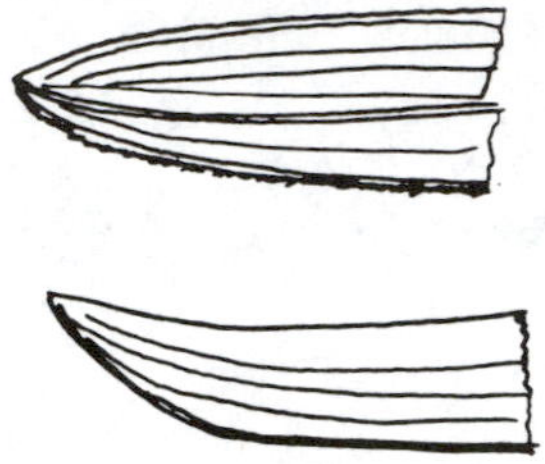

Figure 54

68a Nerves faint; lemmas usually with golden or purplish transverse bands below the tip; plants usually on salty or alkaline soil
. . . . 12. (p. 61) *Puccinellia*

68b Nerves conspicuous; lemmas green or with a purple band below the tip; plants of wet ground or water . . . .
. . . . 13. (p. 62) *Torreyochloa*

69a Plants slender, annual, with shallow roots and hair-like leaves; culms usually less than 20 cm tall; stamens usually 1; florets cleistogamous . . . .
. . . . 8. (p. 56) *Vulpia*

69b Plants perennial, with hard bases; leaf blades flat or rolled or folded, not hair-

like; culms mostly 30 cm or more tall; stamens 3; florets opening at time of pollination . . . .
. . . . 9. (p. 57) *Festuca*

KEY VI

Spikelets Disarticulating Below the Glumes, Falling Separately or with Attached Structures, or in Clusters, or Remaining on the Plants, Only Some Parts Shed

1a Spikelets disarticulating from the inflorescence singly, without attached pedicels, rachis internodes, or bristle-like branches 2

1b Spikelets either remaining on the plant permanently, or falling with attached pedicels, rachis internodes, or bristles, or falling in clusters of 2 to several 39

2a Spikelets large, strongly laterally flattened and keeled, many flowered, 1-1.5 cm wide; glumes much shorter than the spikelet . . . .
. . . . 74. (p. 116) *Uniola*

2b Spikelets smaller, less than 1 cm wide; florets 1 to several; glumes short or as long as the spikelet. 3

3a Foliage covered with sticky hairs and with a strong odor of molasses . . . .
. . . . 109. (p. 153) *Melinis*

3b Foliage not sticky-haired, without molasses odor 4

4a Inflorescence densely silky, the spikelets

concealed by the long hairs borne on the glumes, pedicels or branches 5

4b Spikelets not concealed by long hairs . . . 8

5a Spikelets laterally compressed, borne in an open panicle on delicate, drooping pedicels; spikelets all alike, not paired; hairs usually rosy pink. . . . .
. . . . 127. (p. 180) *Rhynchelytrum*

5b Spikelets dorsally compressed, paired, borne in long racemes; hairs white or tan. 6

6a First glume minute or absent; fertile floret hard, dark brown, as long as the awnless spikelet . . . .
. . . . 124. (p. 178) *Digitaria*

6b First and second glumes equal, as long as the spikelet; fertile floret delicate, hidden between the glumes; spikelets awned or awnless 7

7a Inflorescence slender, cylindrical, the individual racemes scarcely separately visible . . . .
. . . . 128. (p. 181) *Imperata*

7b Inflorescence a fan-shaped group of racemes. . . . .
. . . . 130. (p. 182) *Miscanthus*

8a Spikelets with 2 or more flowers 9

8b Spikelets with a single flower, sometimes with sterile or reduced florets below it . . .
. . . . 13

9a Sheaths with united edges; spikelets

drooping, on weak pedicels . . . .
. . . . 57. (p. 97) *Melica*

9b   Sheaths with overlapping edges; spikelets
     not drooping 10

10a Glumes much shorter than the spikelet;
     florets 3 or more . . . .
     . . . . 16. (p. 64) *Sclerochloa*

10b Glumes longer than the florets 11

11a Glumes equal in length and width 12

11b First glume narrow; second glume several
     times as wide, usually blunt. Fig. 55 . . . . .
     . . . . 29. (p. 77) *Sphenopholis*

Figure 55

12a Glumes hairy; foliage softly velvety;
     florets 2, the upper staminate . . . .
     . . . . 31. (p. 79) *Holcus*

12b Glumes not hairy; foliage not velvety;
     florets several, all perfect-flowered . . . . .
     . . . . 70. (p. 113) *Schismus*

13a Glumas absent or so minute as to appear
     absent; mostly aquatic or marsh
     grasses 14

13b Glumes present, at least the second well
     developed 17

14a Spikelets strongly flattened,
     perfect-flowered; plants growing on damp
     soil or in forests. Fig. 56 . . . .
     . . . . 3. (p. 45) *Leersia*

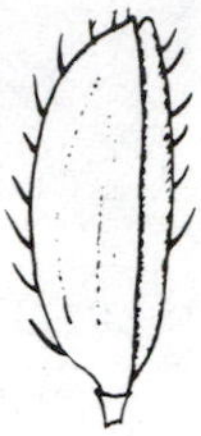

Figure 56

14b Spikelets not strongly flattened; flowers
     unisexual; plants aquatic 15

15a Plants submerged except for the floating
     upper leaves; spikelets few, in small
     clusters . . . .
     . . . . 6. (p. 47) *Hydrochloa*

15b Plants tall, emerging from the water;
     inflorescences large and with many
     spikelets. 16

16a Inflorescence with erect, awned pistillate
     spikelets at its tip and drooping, awnless
     staminate spikelets on the lower branches;
     annual plants. . . . .
     . . . . 4. (p. 46) *Zizania*

16b Inflorescence with staminate and pistillate
     spikelets intermixed on the same
     branches; perennial with rhizomes. . . . . .
     . . . . 5. (p. 46) *Zizaniopsis*

17a Spikelets laterally compressed; no sterile
     florets below the fertile one 18

17b Spikelets dorsally compressed; bracts
     below the fertile floret either 2 (second
     glume and sterile lemma) or 3 (short first

glume, long second glume and sterile lemma) 26

18a Inflorescence a single erect spike or raceme; cultivated grasses 60

18b Inflorescence an open or spikelike panicle or a group of 1-sided spikes; plants wild or cultivated. 19

19a Inflorescence of 1-sided spikes. Fig. 57 20

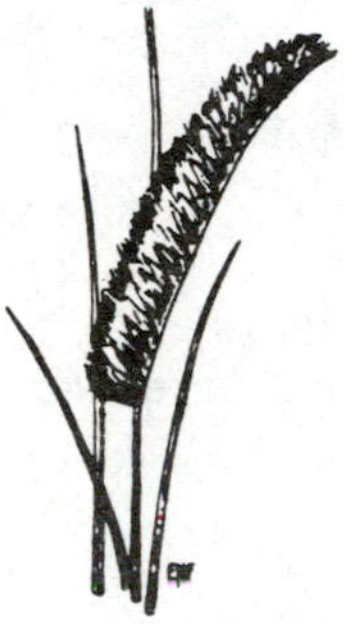

Figure 57

19b Inflorescence an open or dense panicle 21

20a Spikelets nearly circular, the equal sac-like glumes concealing the awnless floret. Fig. 58 . . . . 32. (p. 79) *Beckmannia*

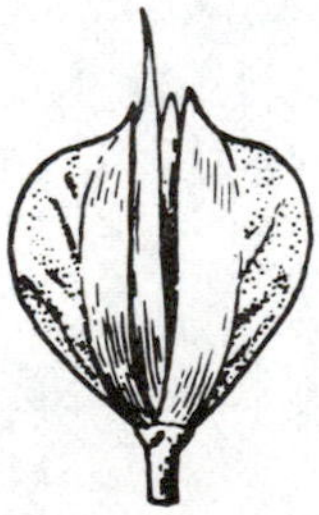

Figure 58

20b Spikelets narrow, the unequal glumes not

concealing the floret . . . . 92. (p. 140) *Spartina*

21a Glumes bearing awns longer than the body . . . . 40. (p. 86) *Polypogon*

21b Glumes not long-awned 22

22a Glumes (actually 2 sterile lemmas) much shorter than the floret; aquatic crop. Fig. 59 . . . . 2. (p. 45) *Oryza*

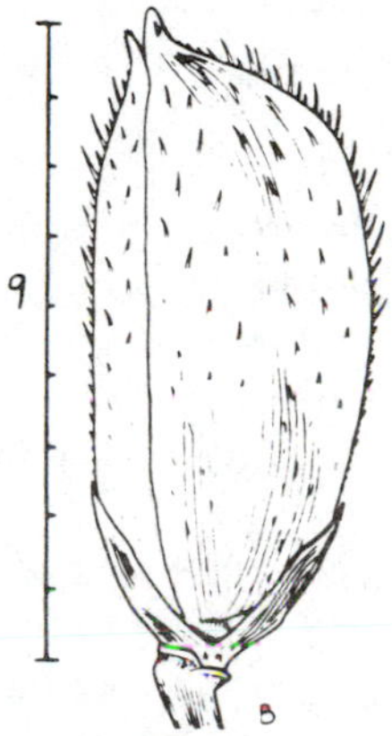

Figure 59

22b Glumes as long as the floret 23

23a Panicle dense, cylindrical, spikelike; rachilla not projecting behind the palea 24

23b Panicle open, loose; rachilla prolonged behind the palea as a minute bristle. Fig. 60 25

Figure 60

24a Glumes acute, joined together near the base; lemma bearing an awn . . . . 44. (p. 87) *Alopecurus*

24b Glumes truncate, with a projecting midrib, not joined at the base; lemma awnless . . . . 45. (p. 88) *Phleum*

25a Awn of lemma minute or absent . . . . 43. (p. 87) *Cinna*

25b Awn of lemma twice as long as the spikelet . . . . 42. (p. 87) *Limnodea*

26a Spikelets interspersed with stiff bristles (sterile branches) longer than the spikelets; inflorescence a dense, pencil-shaped panicle. Fig. 44 . . . . 121. (p. 173) *Setaria*

26b Spikelets not interspersed with sterile bristles, all the branches bearing spikelets. 27

27a Spikelets arranged in 2 or 4 definite rows along the lower sides of 1 or more un-branched racemes 28

27b Spikelets more or less randomly arranged along the simple or divided branches of a panicle. 34

28a Spikelets awned; creeping grasses of warm climates . . . . 116. (p. 170) *Oplismenus*

28b Spikelets awnless 29

29a Spikelets with a hardened, projecting, cup-shaped structure (first glume) at the base. Fig. 61 . . . . 114. (p. 169) *Eriochloa*

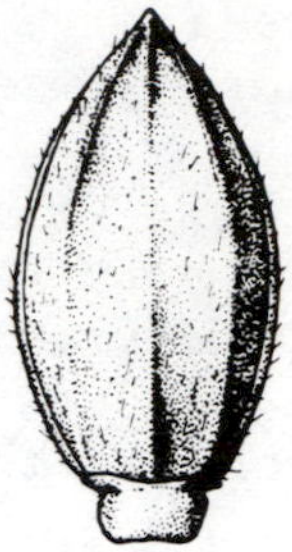

Figure 61

29b Spikelets without a projecting basal knob 30

30a Fertile lemma hard and stiff, its margin rolled in. 31

30b Fertile lemma soft and flexible, its margins not rolled in, exposed and thin . . . . 124. (p. 178) *Digitaria*

31a First glume absent or less than one-tenth as long as the spikelet. 32

31b First glume present, at least one-fourth as long as the spikelet. 33

32a Spikelets placed with the back (convex side) of the fertile lemma toward the midrib of the rachis. Fig 62 . . . . 112. (p. 164) *Paspalum*

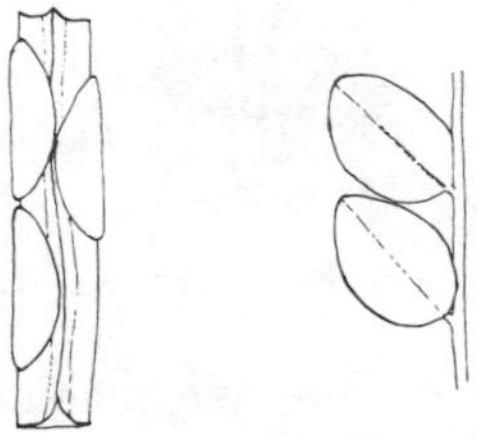

Figure 62

32b Spikelets placed with the back of the fertile lemma turned away from the midrib of the rachis . . . . 125. (p. 179) *Axonopus*

33a Spikelets placed with the first glume toward the midrib of the rachis and the back of the fertile lemma away from it. . . . . 113. (p. 167) *Brachiaria*

33b Spikelets placed with the first glume away from the midrib of the rachis and the fertile lemma toward it; fertile lemma cross-wrinkled. Fig. 63 . . . . 111. (p. 163) *Paspalidium*

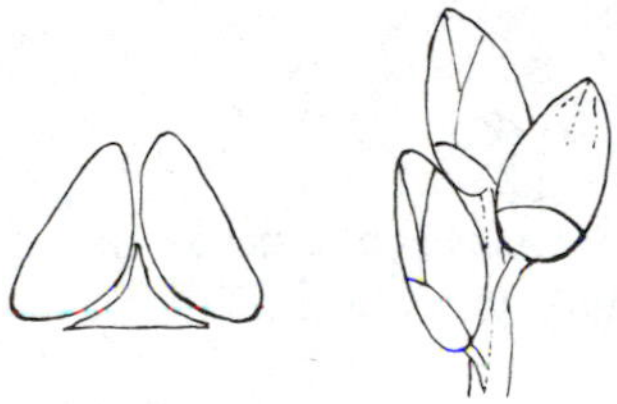

Figure 63

34a Second glume or sterile lemma awned (the awn sometimes reduced to an abrupt point); fertile lemma pointed. Fig. 64 . . . . 117. (p. 170) *Echinochloa*

Figure 64

34b Glumes and sterile lemma awnless 35

35a Second glume saclike, swollen, the spikelet hence lopsided; fertile floret about half as long as the sterile lemma; swamp plants . . . . 118. (p. 171) *Sacciolepis*

35b Second glume not swollen at the base; fertile floret about as long as the spikelet 36

36a First glume at least one-fourth as long as the spikelet (3 bracts present below the fertile floret). Fig. 65 . . . . 110. (p. 154) *Panicum*

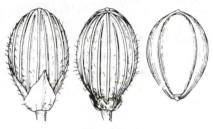

Figure 65

36b First glume absent or minute, less than one-tenth as long as the spikelet 37

37a Spikelets ovoid, blunt, covered with spreading hairs . . . . 115. (p. 169) *Anthaenantia*

37b Spikelets lanceolate, acute, not covered with spreading hairs. 38

38a Panicle narrow, with ascending branches and short-pedicellate spikelets; aerial spikelets sterile; plants producing enlarged underground spikelets at the tips of slender rhizomes; sandy lands of the Atlantic Coastal Plain . . . . 119. (p. 172) *Amphicarpum*

38b Panicle large, open, dome-shaped, the

spikelets borne on elongated, wiry slender pedicels; no underground spikelets . . . . .
. . . . 126. (p. 180) *Leptoloma*

39a   All spikelets remaining on the plant, not shed as units or groups. 40

39b   Spikelets or spikelet groups falling from the plant 41

40a   Lemmas 3 nerved, not toothed; glumes and later the lemmas falling from the rachilla, the paleas usually remaining . . .
. . . . 76. (p. 117) *Eragrostis*

40b   Lemmas many-nerved, toothed at the apex; entire spikelets remaining on the plants; California only . . . .
. . . . 106. (p. 152) *Orcuttia*

41a   Low, creeping stoloniferous grass of the Great Plains; pistillate spikelets borne in hardened, detachable axillary bead-like structures at the base of the plants . . . . . .
. . . . 100. (p. 148) *Buchloë*

41b   Plants of various habits; spikelets borne in open inflorescences at the tips of the culms; spikelets, or some of them, perfect-flowered 42

42a   Inflorescence a dense, cylindrical bristly panicle, the spikelets falling in groups of 1 to several, surrounded by an attached circle of stiff, elongated bristles (sterile branches) . . . .
. . . . 122. (p. 176) *Pennisetum*

42b   Inflorescence various, but never with sterile, bristle-like branches 43

43a   Inflorescence an elongated raceme of numerous short, 1-sided spikes that drop from the rachis at maturity. Fig. 66 . . . . .
. . . . 99. (p. 146) *Bouteloua*

Figure 66

43b   Inflorescence not made up of deciduous 1-sided spikes. 44

44a   Inflorescence a single spike or rame on each peduncle (sometimes these are grouped together) 45

44b   Inflorescence of 2 to many branches on one peduncle 61

45a   Spikelets awnless, fitting into hollows of a jointed, thick rachis. 46

45b   Spikelets awned or awnless, not sunken into a hollow rachis 48

46a   Spikelets all sunken into one side of an erect, corky, flattened rachis; stoloniferous plants with strongly keeled sheaths and subopposite leaves . . . .
. . . . 120. (p. 172) *Stenotaphrum*

46b Spikelets borne on both sides of a more or less cylindrical rachis; plants not stoloniferous, the sheaths not strongly keeled and leaves plainly alternate; rachis readily disarticulating 47

47a Spikelets paired at each node of the rachis, one sessile and perfect-flowered, the other pedicellate and reduced to a rudiment. Fig. 67. . . . .
. . . . 139. (p. 189) *Coelorachis*

Figure 67

47b Spikelets solitary at each node of the rachis, all alike; low, spreading seashore plants . . . .
. . . . 65. (p. 110) *Parapholis*

48a Rachis breaking up into individual segments at maturity, each bearing 1-3 spikelets. 49

48b Rachis remaining intact 58

49a Spikelets 1 at each node of the rachis, falling attached to the upwardly-thickened internode. . . . .
. . . . 50. (p. 92) *Aegilops*

49b Spikelets 2-3 at each node of the flat rachis. 50

50a Spikelets 2 at each node, one sessile and perfect-flowered, the other pedicellate and staminate or sterile, the pair falling together; fertile floret 1 51

50b Spikelets 2-3 per node, all sessile or subsessile; fertile florets 1 to several. . . 55

51a Sessile spikelets round, blackish, rough-pitted, 1-2 mm in diameter, awnless . . . .
. . . . 140. (p. 189) *Hackelochloa*

51b Sessile spikelets not round, awned or awnless 52

52a Spikelets awnless 53

52b Spikelets awned 54

53a Spikelets pointed; both spikelets of each pair similar, hairy; tall erect grasses . . . . .
. . . . 134. (p. 186). *Elyonurus*

53b Spikelets truncate, the apex notched; only the sessile spikelet developed, accompanied by a flattened pedicel bearing a rudiment; low, creeping lawn grass. Fig. 68. . . . .
. . . . 141. (p. 190) *Eremochloa*

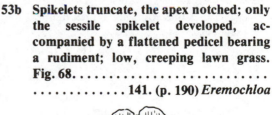

Figure 68

54a Awns hairy, twisted, 5-12 cm long; perfect-flowered spikelet looks like a *Stipa* floret . . . . 142. (p. 190) *Heteropogon*

54b Awn not hairy, less than 2 cm long; perfect-flowered spikelet dorsally flattened, not *Stipa*-like . . . . 132. (p. 183) *Schizachyrium*

55a Spikelets 1 flowered, 3 borne at each node . . . . 53. (p. 95) *Hordeum*

55b Spikelets several-flowered, 2 or 3 borne at each node 56

56a Spike with 2 spikelets at some lower nodes, but 1 spikelet at upper nodes. . . . . 54. (p. 96) *X Agrohordeum*

56b Spike bearing 2-3 spikelets per node throughout 57

57a Spikelets 3 per node, each with 1-2 flowers; sterile hybrids . . . . 55. (p. 97) *X Elyhordeum*

57b Spikelets 2 per node, each several-flowered; plants producing good seed. . . . . 56. (p. 97) *Sitanion*

58a Spikelets at each node dropping as a group from the zigzag rachis . . . . 102. (p. 149) *Hilaria*

58b Spikelets 2 at each node of the straight rachis, of different structure, 1 or both remaining on the rachis permanently. . . . 59

59a Perfect-flowered spikelets cylindrical,

long-awned, falling with the hairy pedicel attached; staminate spikelets awnless, remaining on the rachis . . . . 143. (p. 191) *Trachypogon*

59b Spikelets awnless, the perfect-flowered ones dorsally compressed, the first glume truncate and notched at the center; pedicellate spikelet reduced to a small rudiment on a flattened pedicel; both spikelets remaining on the persistent rachis. Fig. 68 . . . . 141. (p. 190) *Eremochloa*

60a Spikelets 3 at each node of the rachis, sessile, awned, each with 2 glumes; cereal grain. Fig. 69. . . . . 53. (p. 95) *Hordeum*

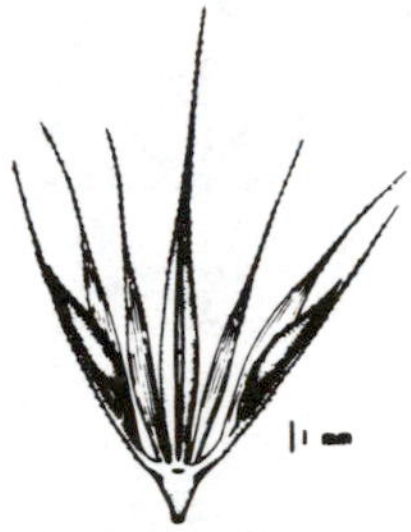

Figure 69

60b Spikelets 1 at each node of a slender raceme, pedicellate, awnless, each with a single glume; lawn grass . . . . 107. (p. 152) *Zoysia*

61a Spikelets falling as groups of 2 or more from the branches of the panicle, which remain on the plant 62

61b Spikelets paired, each pair falling attached to its rachis internode, the branches thus disintegrating. 63

62a Spikelets 2 in a group, both 1 flowered,

similar, one staminate and one perfect-flowered ....
.... 87. (p. 136) *Lycurus*

62b Spikelets more than 2 in each group, one fertile, single-flowered, awned, the others sterile, awnless, with numerous florets. . .
.... 22. (p. 74) *Lamarckia*

63a Each branch (rame) consisting of many consecutive paired or rarely solitary spikelets, the rames attached directly to the central rachis of the inflorescence. . 64

63b Each branch (rame) consisting of 1-3 (rarely up to 7) pairs of spikelets, the rames borne on the branches of open or dense panicles 68

64a Spikelets awnless, or the awn so small as to be hidden within the glumes....... 65

64b Spikelets with a well-developed exserted awn.......................... 67

65a Giant cultivated grasses, the solid culms 2-5 cm thick; plants 2-5 m tall, occasionally bearing large, plume-like silky white panicles.........................
.... 131. (p. 182) *Saccharum*

65b Low creeping grasses; inflorescences of few rames, not large and silky; wild plants 66

66a Spikelets paired, equal, one sessile and

one pedicellate at each node of the rachis.
.... 135. (p. 186) *Microstegium*

66b Spikelets solitary, sessile at each node of the rachis, some of the lower ones accompanied by a bristle-like sterile pedicel
.... 136. (p. 186) *Arthraxon*

67a Sessile spikelet of each pair perfect-flowered, the stalked spikelet reduced in size, awnless, either staminate or rudimentary. Fig. 70 ....
.... 133. (p. 184) *Andropogon*

Figure 70

67b Both spikelets of each pair equal in size and both with perfect flowers and awns. .
.... 129. (p. 181) *Erianthus*

68a Each hard, fertile, sessile, awned spikelet accompanied by a soft, stalked, awnless, staminate one ....
.... 137. (p. 187) *Sorghum*

68b Each fertile, sessile spikelet accompanied by a hairy rachis internode and a similar pedicel which is sterile, lacking a spikelet
.... 138. (p. 188) *Sorghastrum*

Pictured Keys to Common American Grasses

Subfamily I. Bambusoideae
BAMBOO SUBFAMILY

Tribe 1. Arundinarieae

1. Arundinaria CANE
Figure 71 . . . . **CANE** *Arundinaria gigantea*
(Walt.) Chapm.

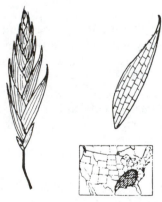

Figure 71

Stems (canes) woody, perennial, reaching as

much as 10 m in height; plants rarely flowering. Cane grows in dense colonies, called cane-brakes, in river bottomlands from Virginia to the Ohio Valley and southward to eastern Texas and Florida. The canes are used for fish poles, basketry, and in a variety of other ways. The young shoots and leaves are eagerly taken by domestic animals. A smaller form of cane, with culms usually less than 2 m tall, is *Arundinaria tecta* (Walt.) Muhl.

This species can be distinguished by the presence of air canals in the rhizomes. These are lacking in *A. gigantea*.

While cane is our only native bamboo, a number of other species are cultivated in mild climates. A few are hardy as far north as New York and St. Louis, and a number are cultivated in the Southeast and in the Pacific Coast States. Species of *Phyllostachys,* with culms D-shaped in cross section, and paired lateral branches, are especially common.

Subfamily II. Oryzoideae

Tribe 2. Oryzeae

2. Oryza RICE
Cultivated aquatic annual; inflorescence a panicle; spikelets very flat, awned or awnless; glumes reduced to a minute cup; floret disarticulating above this, carrying with it 2 sterile lemmas about 1/3 as long as the fertile floret. Fig. 72 . . . . **RICE** *Oryza sativa* **L.**

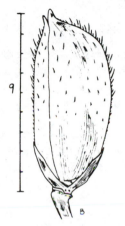

Figure 72

Annual; plants aquatic, stout, 1-2 m tall; panicles drooping, the very flat spikelets hairy, awned or awnless. The structures which look like glumes are really sterile lemmas; the true glumes are the minute ridges which are left behind at the summit of the pedicel when the floret drops. Rice is one of the principal food crops of the world but its culture is restricted to moist or irrigated regions with warm temperate or tropical climates. In the United States, it is grown only in the lower Mississippi Basin and in Florida and California.

3. Leersia CUT GRASS
Rhizomatous perennials; inflorescence a panicle; spikelets very flat, awnless; glumes reduced to a minute cup; floret disarticulating above this.

1a **Sheaths strongly downwardly scabrous; rhizomes slender, with exposed internodes; lower panicle branches whorled. Fig. 73 . . . . CUT GRASS** *Leersia oryzoides* **(L.) Sw.**

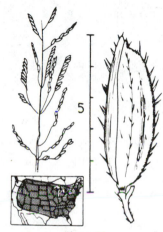

Figure 73

Perennial; culms up to 1.5 m long, weak and often sprawling; rhizomes long and slender; leaf sheaths and margins of the blades armed with very sharp minute spines which can scratch severely. The plants are very rough to the touch and cling readily to clothing. The glumes of species of *Leersia* are the minute cuplike structures from which the floret drops. Cut grass often forms dense "jungles" along streams or around ponds or in marshes. July-October.

Leersia lenticularis Michx. (CATCHFLY GRASS) is similar, but has broadly oval spikelets, 3-4 mm wide and 4-5 mm long, arranged in neat overlapping rows. Wet ground and swamps, Mississippi Valley and southeastern states.

1b Sheaths smooth or nearly so; rhizomes short and thick, densely covered with scales; lower panicle branches borne singly. Fig. 74 . . . .
. . . . **WHITE GRASS** *Leersia virginica* **Willd.**

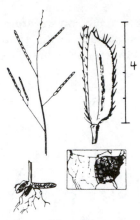

Figure 74

Perennial; culms 50-120 cm tall, weak and slender; panicle 10-20 cm long, with a few simple branches, the spikelets lying closely parallel to them. Some of the smaller panicles may be hidden in the sheaths. Leaf blades yellowish-green. Damp woods and thickets; mud flats. July-October.

Tribe 3. Zizanieae

4. Zizania WILD RICE
Tall aquatic annuals; pistillate spikelets all at the upper tip of the panicle, forming an erect brush, their awns several times as long as the lemma; staminate spikelets awnless, drooping, all on spreading lower branches of the panicle. Fig. 75 . . . .
. . . . **WILD RICE** *Zizania aquatica* **L.**

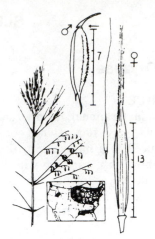

Figure 75

Annual; culms stout, 2-3 m or more tall; panicles 30-50 cm long, open and pyramidal. The spikelets consist of single florets, which disarticulate from minute cuplike structures which are the vestiges of glumes. Wild rice was an important food plant for the American Indians, who threshed the standing plants into canoes. It still furnishes some food for human beings and the grain can occasionally be purchased in stores. In nature, wild rice is an important producer of food for waterfowl. Shallow water, ditches, ponds, streams and marshes. July-September.

5. Zizaniopsis SOUTHERN WILD RICE
Tall aquatic perennial; pistillate spikelets intermixed with the staminate ones on the same branches, awn-tipped; caryopsis oblong, free from the floret, with a short stiff persistent style. Fig. 76 . . . .
. . . . **SOUTHERN WILD RICE** *Zizaniopsis miliacea* **(Michx.) Doell**

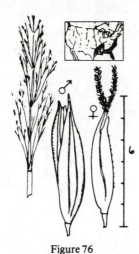

Figure 76

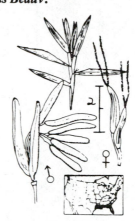

Figure 77

spikelets in separate small panicles. Fig. 77 . . . .
. . . . **WATER GRASS** *Hydrochloa caroliniensis Beauv.*

Perennial; plants 1-3 m tall or taller, the culms arising from stout rhizomes; panicles nodding, 30-50 cm long. Leaves very scabrous on the margins. The spikelets consist of single florets, which disarticulate from the vestigial glumes, as in the previous species. The staminate ones have 6 stamens instead of the usual 3. Marshes and along streams. May-June.

6. Hydrochloa

Immersed aquatics, only the upper leaves floating on the water; staminate and pistillate

Perennial; culms slender, weak, up to 1 m long, floating in water, the upper leaves on the surface. The staminate spikelets are borne at the tips of branches and the pistillate ones in the axils of leaves. Neither type has evident glumes, and both have only a single floret. Ponds and slow streams. Furnishes some feed for livestock. Blooming apparently rare. June-August.

Subfamily III. Pooideae
POOID SUBFAMILY

Tribe 4. Poeae

7. Bromus BROME GRASSES

1a Spikelets strongly laterally flattened, 2-4 cm long; lemmas V-shaped in cross section. 2

1b Spikelets not strongly flattened, round in

cross section before flowering; lemmas rounded on the back 3

2a Lemmas awnless or with a short awn less than 2 mm long. Fig. 78. . . . .
. . . . **RESCUE GRASS** *Bromus catharticus Vahl*

Figure 78

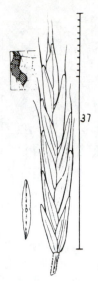

Figure 79

Annual; tufted, culms up to 100 cm long, erect or spreading; leaf sheaths and blades glabrous or hairy, dark green; panicles open, up to 20 cm long; spikelets 2-3 cm long, with 6-12 florets; lemmas glabrous or rarely hairy, about 1.5 cm long, much flattened and closely overlapping. Rescue grass got its name from its winter annual habit, which makes it one of the earliest forage grasses in the South. It is planted in the fall for winter and spring pasture, but in many areas it has escaped from cultivation and is regarded as a wild plant. With good moisture, it makes lush, highly palatable forage. Heavy rich soil, bottomlands. Native to South America. March-June.

Previously called *B. unioloides* H.B.K., a name now shown to be antedated by *B. catharticus*.

2b **Lemmas bearing awns 5-15 mm long. Fig. 79. . . . . MOUNTAIN BROME *Bromus carinatus* H. & A.**

Annual or biennial; tufted; plants 50-100 cm tall or taller, vigorous and leafy; panicles 15-30 cm long, with spreading or drooping branches; sheaths and leaf blades smooth or hairy; blades ranging from narrow and involute to broad and flat. A number of closely related and intergrading plants, sometimes recognized as separate species, are included here. These plants are common on open ground and in thin woods in the western states and furnish a good deal of range forage. The foliage and the seed heads are eaten, the latter furnishing a good fattening ration for lambs. The seed of these plants is now available in commerce and the plants are used for range revegetation in the West. March-June.

3a **Lemmas broad, rounded or tapered to the apex, the lateral teeth at the base of the awn blunt or united. Fig. 80A 4**

3b **Lemmas narrow, with a hard sharp callus and long, sharp lateral teeth at the base of the awn. Fig. 80B. 16**

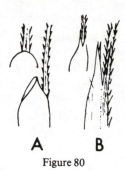

Figure 80

4a First glume 1 nerved, narrowly lanceolate 5

4b First glume 3-5 nerved, ovate or elliptical 9

5a Plants tufted, without rhizomes; panicles mostly drooping; lemmas bearing well-developed awns 6

5b Plants bearing rhizomes; panicles erect, with ascending branches; lemmas awnless or with very short awns, 1-2 mm long. Fig. 81 . . . . **SMOOTH BROME** *Bromus inermis Leyss.*

Figure 81

Perennial; 50-100 cm tall; panicles 10-20 cm long. Smooth brome, introduced from Eurasia, is one of our most successful forage grasses, and has been very widely planted in the United States for pasture and hay production. It frequently escapes to roadsides, ditches, and moist wooded areas. June-August.

Bromus pumpellianus Scribn. is a closely related species, native to the western states from the Black Hills to Colorado and Alaska. It has rhizomes but the lemmas are hairy. It hybridizes with *B. inermis*.

6a Lemmas pubescent along the margins and lower part of the back, the central portion glabrous. Fig. 82 7

Figure 82

6b Lemmas pubescent across the back 8

7a Ligule 3-5 mm long; awns 5 mm or more long; plants of the western states. Fig. 83 . . . . *Bromus vulgaris* (Hook.) Shear

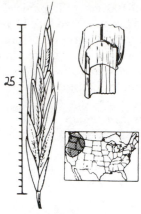

Figure 83

Perennial; tufted; plants slender; culms 80-120 cm tall; ligules prominent; leaf blades up to 12 mm wide; panicles drooping, 10-15 cm long; spikelets about 2.5 cm long; lemmas usually 8-10 mm long, hairy on the margins, glabrous or nearly so on the back; awns 5-8 mm long. Moist rocky woods and canyons. Forms with nearly glabrous foliage and lemmas are known. July-August.

7b Ligule about 1 mm long; awn 3-5 mm long; plants widespread. Fig. 84 . . . . *Bromus ciliatus* L.

Figure 84

Perennial; tufted; plants 70-120 cm tall; panicles 15-25 cm long, drooping. Leaf sheaths glabrous or somewhat hairy; blades smooth or hairy, up to 1 cm wide. This species has handsome fringed spikelets. It is one of the most widespread of the native woodland bromes in moist rocky or alluvial woods. It provides excellent forage in the western states. July-August.

8a Culms with 3-7 nodes; sheaths without auricles. Fig. 85 . . . . *Bromus pubescens* L.

Figure 85

Perennial; tufted, in small clumps; culms erect or leaning; plants usually 60-150 cm tall; panicles open, drooping, usually 15-20 cm long; leaf sheaths shorter than the internodes, hairy or rarely glabrous; leaf blades 5-17 mm wide; lemmas rather uniformly hairy across the back. This is the commonest woodland brome in the eastern United States. It is to be found in nearly every moist woods. Called *B. purgans* in recent manuals. June-July.

Forma *glabriflorus* is a form of this species which has glabrous lemmas. It may be distinguished from other similar woodland bromes by the anthers, which are 3-4.5 mm long.

8b **Culms with 10-20 nodes, the sheaths longer than the internodes, bearing pointed appendages (auricles) at the throat. Fig. 86 . . . .**
. . . . ***Bromus purgans* L.**

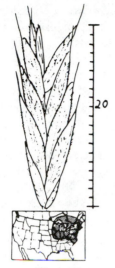

Figure 86

Perennial; tufted; culms up to 2 m tall; panicles usually 15-25 cm long, rather dense. Leaf sheaths longer than the internodes, hence overlapping. This species blooms several weeks later than *B. pubescens*. These two species frequently grow together. The sheaths are often covered with dense grayish wool. Alluvial bottomlands; prairies. Called *B. latiglumis* in most manuals. July-September.

9a **Panicle open, pyramid-shaped, erect or drooping 10**

9b **Panicle dense, ovoid, erect, with short branches and overlapping spikelets . . . 15**

10a **Lemmas glabrous or scabrous 11**

10b **Lemmas heavily pubescent. 14**

11a **Lemmas bearing awns, not broad or inflated 12**

11b **Lemmas awnless or with minute awn tips, very broad and inflated. Fig. 87 . . . .**
. . . . **RATTLESNAKE CHESS *Bromus brizaeformis* F. & M.**

Figure 87

Annual; tufted; plants 30-60 cm tall; panicles drooping, 5-15 cm long. The odd, inflated spikelets of this species look much like the rattlers of a rattlesnake. It is sometimes planted for ornament and is occasionally found naturalized in fields and waste ground in the western states and elsewhere. Introduced from Europe. June-August.

12a **Lemmas overlapping; rachilla not exposed; awns well developed; upper sheaths pubescent 13**

12b **Margins of lemmas rolling inward at maturity, exposing the rachilla; awns short, kinked; upper sheaths glabrous. Fig. 88 . . . .**
. . . . **CHESS, CHEAT *Bromus secalinus* L.**

Figure 88

Annual; tufted; plants 30-60 cm tall; panicles 7-12 cm long. Chess is a common weed of road-sides and grainfields. The quickly-maturing seeds may be harvested with wheat or other small grains and replanted elsewhere. It is particularly common where wheat is grown. Introduced from Europe. Old superstition claimed that cheat came from degenerate small grains, hence the name. May-July.

13a   **Mature spikelets 3-5 mm wide; lower sheaths densely woolly with tangled or matted soft hairs. Fig. 89. . . . .** . . . .**JAPANESE BROME**
Bromus japonicus **Thunb.**

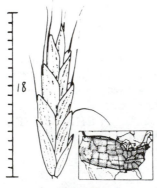

Figure 89

Annual; tufted; plants 40-70 cm tall; panicles drooping, with delicate, flexuous branches. The awns may be straight or bent, depending upon their moisture content. Japanese brome was introduced from the Old World, and is now a very widespread weed of roadsides, fields, and waste ground. May-August.

Bromus arvensis L., FIELD BROME, is similar to this species but has more slender spikelets, and anthers 3-4 mm long, those of *B. japonicus* being under 2 mm long. Field brome is rare in this country, but has been promoted by a few seedsmen in recent years.

13b.   **Mature spikelets 5-8 mm wide; lower sheaths covered with straight spreading stiff hairs. Fig. 90 . . . .** . . . .***Bromus commutatus* Schrad.**

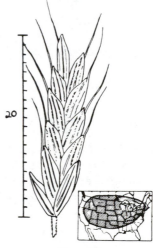

Figure 90

Annual; tufted; plants usually 30-100 cm tall, freely branching from the base; foliage hairy; panicles open, pyramidal, usually 5-15 cm long. In plants growing on sterile dry soil, the inflorescence may be reduced to a raceme of a few spikelets. Such plants closely resemble *B. racemosus* (Fig. 94). This species is closely related to *B. secalinus,* from which it differs in the greater hairiness of the foliage and more overlapping florets. Along with *B. japonicus* and *B. secalinus,* this species is a widespread weed of fields and waste places. It is particularly common in the eastern and far western states, but apparently somewhat rare in the Midwest. Introduced from Europe. June-July.

14a **Leaf blades 2-4 mm wide, sparsely hairy; from North Dakota to western Texas and westward. Fig. 91** . . . . *Bromus anomalus* **Rupr.**

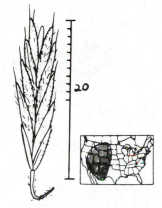

Figure 91

Perennial; tufted; culms slender, 30-60 cm tall; nodes hairy; sheaths somewhat hairy or glabrous; panicles small, drooping, usually 10 cm long or shorter; spikelets few, drooping, densely hairy; first glume with 3 nerves, the second with 5; lemmas about 12 mm long, the awns 2-4 mm long. *Bromus anomalus* is widespread and common in the Rocky Mountain region, where it is regarded as a very valuable forage grass for all kinds of domestic livestock and for wild grazing animals. It grows in aspen, spruce, and pine forests and on open ground in meadows and parks, at elevations up to 3000 m. July-September.

Var. *lanitipes* (Shear) Hitchc. has woolly sheaths.

14b **Leaf blades 5-10 mm wide, densely hairy; Minnesota and Iowa eastward.**

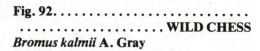

Fig. 92. . . . . **WILD CHESS** *Bromus kalmii* **A. Gray**

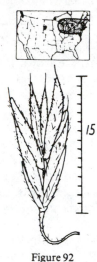

Figure 92

Perennial; tufted; plants 50-100 cm tall; panicles small, 5-10 cm long, drooping. The lemmas are very conspicuously hairy. This and the preceding species are the only native perennial woodland species with 3 nerved first glumes. Roadsides and open woods. July-August.

15a **Lemmas hairy. Fig. 93.** . . . . *Bromus mollis* **L.**

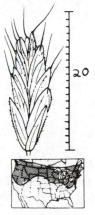

Figure 93

Annual; tufted; plants 20-80 cm tall; leaf sheaths and blades softly hairy; panicles stiff, dense, and erect, 5-10 cm long; glumes and lemmas hairy; first glume with 3 or 5 nerves, the second with 5 or 7; lemmas soft, with 7 nerves, usually 7-9 mm long. This weedy species was introduced from Europe. It is found occasionally in the eastern and middle-western states, but has become very abundant in the Pacific coastal states. It provides short-season spring forage, but because of its shallow roots and annual habit, does not effectively protect the soil from erosion and is regarded as much inferior to the perennial grasses which it replaces in overgrazed areas. *Bromus mollis* hybridizes with *B. racemosus*. April-June.

15b Lemmas glabrous. Fig. 94 . . . .
. . . . *Bromus racemosus* **L.**

Figure 94

Annual; tufted; 20-80 cm tall; panicle stiff, dense, erect. *Bromus racemosus* closely resembles *B. mollis* and hybridizes with it. It is much less common in the eastern states, however. Starved plants of *B. commutatus* resemble *B. racemosus* but have more open panicles. This is a weedy annual of open ground, introduced from Europe. Low value, short-season spring forage. Spring.

16a Panicle open, the branches spreading or drooping **17**

16b Panicle dense, erect. Fig. 95 . . . .
. . . . *Bromus rubens* **L.**

Figure 95

Annual; tufted; plants 15-40 cm tall; panicles 4-8 cm long. The little reddish, bushy panicles look like ragged bristle brushes. Common in the intermountain region and Pacific coastal states, on open dry ground. The awns may injure livestock by piercing the facial tissues. Poor, scanty feed. Introduced from the Mediterranean area. March-June.

17a Second glume at least 12 mm long; lemmas glabrous or sparsely stiff-hairy; awns 2-5 cm long **18**

17b Second glume 10 mm long or shorter; lemmas usually softly pubescent; awns 1-2 cm long. Fig. 96 . . . .
. . . . **DOWNY BROME**
Bromus tectorum **L.**

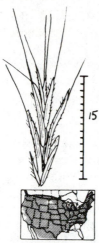

Figure 96

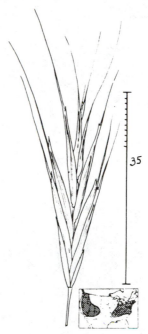

Figure 97

Annual; tufted; plants 30-60 cm tall; panicles drooping, 5-15 cm long. The drooping, reddish panicles are rather ornamental but the plants make poor, sparse feed and the sharp-pointed callus of the lemmas or the awns may penetrate the facial tissues of grazing animals. Of all the weedy European bromes, this is the commonest and most widespread. May-June. *Melica smithii* (Fig. 201) might be keyed here by the unwary. *M. smithii* is a glabrous perennial species of woodlands in cold regions.

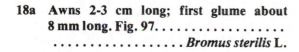

18a **Awns 2-3 cm long; first glume about 8 mm long. Fig. 97**. . . . . ***Bromus sterilis* L.**

Annual; tufted; culms erect; plants 50-100 cm tall; leaf sheaths softly hairy or nearly smooth; ligules prominent, membranous, with a lacerated edge; leaf blades soft, sparsely hairy or nearly glabrous; panicles 10-20 cm long, open, with rather stiffly spreading or drooping branches; spikelets 2.5-3.5 cm long, with 6-10 florets; lemmas 17-20 mm long, scabrous or stiff-pubescent; lateral teeth of the lemma about 2 mm long. The stiff florets, provided with sharp calluses and stiff barbed awns, penetrate the facial tissues of grazing animals. Since the plants are shallow-rooted annuals, they soon dry up and their forage value becomes very low. While very widespread in the United States, *B. sterilis* is nowhere particularly common. Introduced from Europe. April-July.

18b **Awns 3-5 cm long; first glume 1.5-2 cm long. Fig. 98** . . . . **RIP-GUT GRASS** ***Bromus diandrus* Roth**

Figure 98

dangerous properties of the florets. Poor, short-season feed. Introduced from Europe; especially common in California; rare in the eastern states. April-August.

Formerly known as *B. rigidus* in U.S. manuals.

8. Vulpia

Small, tufted annual grasses; inflorescence a panicle, usually narrow and rather few-flowered; spikelets with short glumes and 3 or more florets; lemmas narrow, faintly 5 nerved, awned from the very tip; florets usually cleistogamous (self pollinating within the closed lemma and palea), the single anther usually placed between the stigmas and sending pollen tubes directly into them; disarticulation above the glumes and between the florets. Formerly placed in the genus *Festuca*. Fig. 99 . . . .
. . . . **SIX WEEKS FESCUE**
Vulpia octoflora (Walt.) Rydb.

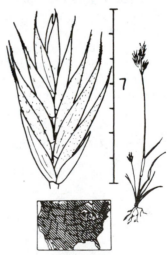

Figure 99

Annual; tufted; plant 40-70 cm tall; leaf sheaths and blades coarsely and sparsely hairy; ligules 3-7 mm long, membranous, with a lacerated margin; panicles dense, with few spikelets, drooping, 7-15 cm long, the lower branches only 1-2 cm long. Some variants have longer lower panicle branches, hence a more open panicle. Spikelets usually with 5-7 florets; glumes glabrous; lemmas scabrous or hairy; lateral teeth 3-5 mm long; awns strong, barbed. Because of the sharp calluses and strong, stiff awns of the lemmas, the florets readily penetrate the soft facial tissues of grazing animals, inflicting bad puncture wounds around the nose, mouth, and eyes. Frequently these become infected, resulting in pink eye, cancer eye, or other disease conditions. The common name, rip-gut grass, arises from these

Annual; tufted, seldom over 20 cm tall. The leaves are borne mostly in a short basal tuft. The plants may be found on poor, usually sandy ground, throughout the United States. Forage value very low. The common name refers to the short life span. This and a number of small species of other genera, with similar growth habits, go by the name of "six weeks grasses." All may provide short-term emergency feed for range livestock after rains. April-July.

About ten other closely related species occur in the United States, especially in the Southwest. They are much less common than the above species.

9. Festuca FESCUE

Tufted perennial grasses; inflorescence a panicle; spikelets several-flowered, disarticulating above the glumes and between the florets; lemmas narrow, tapering into an awn or awn-tip. Stamens 3; flowers opening at pollination time.

1a **Leaf blades flat, soft, the larger ones more than 3 mm wide; lemmas awnless or nearly so** 2

1b **Leaf blades rolled or folded, firm, less than 3 mm wide; lemmas awned or awnless** 3

2a **Spikelets usually less than 10 mm long, with 5 or fewer florets. Fig. 100** . . . .
. . . . **NODDING FESCUE**
Festuca obtusa **Biehler**

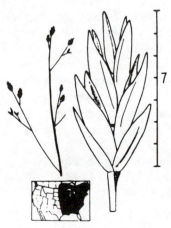

Figure 100

Perennial; tufted; plants 50-100 cm tall; panicles usually 15-20 cm long. Nodding fescue is a species found in forests of the eastern half of the United States. The plants grow in small clumps, with somewhat spreading culms and drooping panicles. The spikelets shatter almost before reaching full size, making it difficult to find complete ones. May-September.

A similar but rarer species is *F. paradoxa* Desv., which has 8-20 spikelets, 4-6 mm wide, on each lower panicle branch.

2b **Spikelets 8-18 mm long, with 8-10 florets. Fig. 101** . . . .
. . . . **MEADOW FESCUE**
Festuca pratensis **Huds.**

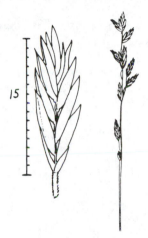

Figure 101

Perennial; tufted; plants 50-120 cm tall; leaf blades 4-8 mm wide; panicle 10-20 cm long, narrow—cylindrical while flowering, but contracted and spikelike afterward. This species was introduced from Europe as a forage plant and has now become widely dispersed in meadows, pastures, roadsides and waste places in the northern states. Forage value good. Also called *F. elatior*. June-July.

ALTA FESCUE OR KENTUCKY 31 *Festuca arundinacea* Schreb. has similar spikelets. It is a coarse, tough grass with elongated leaf blades with coarse ridges on the upper surface. It is very commonly used as a forage grass in the southeastern states, especially south of the Ohio River and often planted for roadside stabilization; also grown under irrigation in the West. At times eating the forage causes lameness and gangrene of the legs in cattle, especially in cold weather (fescue foot disease). This species makes a coarse, bumpy, tough lawn.

3a   Lemmas awnless **4**

3b   Lemmas bearing awns 2 mm or more long **6**

4a   Ligules very short, less than 1 mm long **5**

4b   Ligules 2-4 mm long. Fig. 102 . . . .
. . . . *Festuca thurberi* **Vasey**

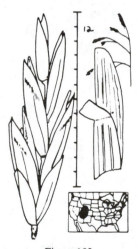

Figure 102

Perennial; densely tufted; culms rather stout, 60-100 cm tall. The involute leaves are scabrous. Panicles 10-15 cm long, the branches separate or paired, as much as 8 cm long, and

bearing spikelets only near their ends. The long ligule is characteristic. Dry rocky slopes, 2500-3500 m elevation. Forage value good. July-August.

5a   Leaves soft, green. Fig. 103 . . . .
. . . . **GREEN FESCUE**
Festuca viridula **Vasey**

Figure 103

Perennial; in dense tufts; culms 50-100 cm tall; panicles open, 10-15 cm long; branches mostly paired, 2-4 cm long; leaf blades soft, folded or rolled; spikelets with 3-6 florets; lemmas 6-8 mm long. The lemmas are softer than those of other species of fescue, and frequently have some purplish coloration. Well-drained soils in the spruce-fir forests near timber line, mountainsides, parks, and meadows. Green fescue is one of the best forage grasses of the Northwest. It is high in palatability and nutritive value, and is eaten throughout the grazing season by all classes of livestock. July-September.

5b   Leaves stiff, bluish. Fig. 104 . . . .
. . . . **ARIZONA FESCUE; PINEGRASS**
Festuca arizonica **Vasey**

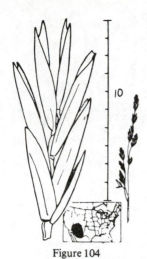

Figure 104

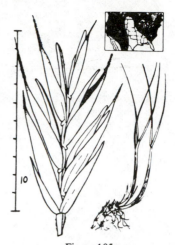

Figure 105

Perennial; densely tufted; culms about 50 cm tall; leaves stiff, pale, rather scabrous; panicles narrow, 8-20 cm long, with one or two spreading branches at the base. Open pine forests, mountains of the Southwest. An important grazing grass, closely related to the more northerly *F. idahoensis* (see Fig. 107). July-August.

Festuca tenuifolia Sibth. commonly called HAIR FESCUE, is a much smaller plant, with spikelets under 5 mm long, and short, spikelike panicles less than 5 cm long. It occurs in lawns and waste places in the eastern states and westward to Illinois; Oregon. Introduced from Europe.

6a New leafy shoots arising within the old sheaths; plants forming dense tufts; old sheaths not reddish nor fibrous 7

**6b New leafy shoots breaking through the bases of the sheaths and spreading at the base, the plants hence forming loose turfs; old basal sheaths reddish brown, finally shredding into brownish threads Fig. 105. . . . .
. . . . RED FESCUE *Festuca rubra* L.**

Perennial; tufted; culms 40-100 cm tall; panicle slender, the short branches ascending. Wet or dry ground; widespread in the eastern United States and in the western mountains. A number of forms of this species are used in lawn mixtures. Chewings fescue and creeping red fescue are among these. Red fescue is native in Europe, Asia, and North Africa as well as North America. Most of the occurrences in the eastern states seem to be introductions, probably in lawn seed mixtures. May-July.

Festuca occidentalis Hook. has somewhat similar culm bases but the awns are as long as the lemmas and the long, spreading panicle branches bear spikelets only near their tips. Northwestern states; northern Michigan and Ontario.

**7a Leaf blades less than half the culm length; panicles less than 10 cm long; plants usually less than 30 cm tall; widespread in the United States. Fig. 106. . . . .
. . . . . SHEEP FESCUE *Festuca ovina* L.**

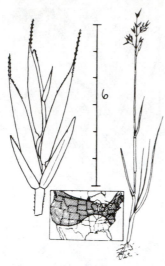

Figure 106

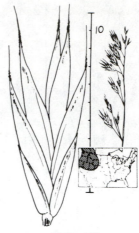

Figure 107

Perennial; tufted; 20-40 cm tall, with a small, narrow panicle. Since the sterile leafy shoots (innovations) arise within the old sheaths, the plants form dense, bumpy tufts, making this species undesirable as a lawn grass. It frequently appears in old neglected lawns. In mass, the plants have a grayish-green color, but they do not turn brown during dry spells, as bluegrass does. Regarded as good forage in the western mountains. Native also to Europe and Asia. May-June.

Var. *brachyphylla* (Schult.) Piper is a dwarf, high altitude form found above timber line in the western mountains, where it is important for grazing. Culms 5-20 cm tall.

7b Leaf blades more than half the culm length; plants 30-100 cm tall; panicles 10-20 cm long; in the western mountains. Fig. 107. . . . .
. . . . BLUEBUNCH FESCUE
Festuca idahoensis **Elmer**

Perennial; forming large tufts, the densely crowded slender culms up to 1 m in height. The panicles are slender, 10-20 cm long, usually with a single longer spreading lower branch. The leaves are rough and glaucous, but not as stiff as those of the closely related Arizona fescue (Fig. 104). Important as a range forage grass. June-August.

10. Leucopoa SPIKE FESCUE
Panicles erect, slender; spikelets unisexual; florets 3-5, the lemmas pointed but awnless, 5 nerved. Fig. 108. . . . .
. . . . **SPIKE FESCUE**
Leucopoa kingii (S. Wats.) Weber

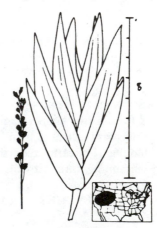

Figure 108

Perennial; tufted or sometimes bearing rhizomes; plants 40-100 cm tall; panicles 7-20 cm long. Spike fescue has the staminate and pistillate spikelets on separate plants (dioecious). While the spikelets are similar, the pistillate ones have well-developed ovaries and abortive anthers, less than 1 mm long, while the staminate ones have large anthers, about 4 mm long, and no ovaries. Grazed by cattle and sheep. Dry mountains, at medium altitudes. Also called *Festuca kingii* and *Hesperochloa kingii*. May-August.

11. Lolium RYEGRASS

The plants of this genus resemble members of the barley tribe (Triticeae) in having symmetrical spikes and have usually been included in that tribe. However, species of *Lolium* cross freely with *Festuca pratensis* and these two genera are closely related. In *Lolium,* the spikelets are placed with one edge against the rachis. Only the exterior glume is developed, except in the spikelet at the very tip of the rachis. Fig. 109 . . . .
. . . . RYEGRASS *Lolium perenne* L.

Figure 109

Winter annual or short-lived perennial; culms 30-60 cm tall; leaves dark green. The lemmas are awnless or nearly so. Ryegrass is much used in lawn seed mixtures, since the seed germinates rapidly and gives a green turf quickly, but the plants tend to die out when hot weather arrives. Lawns, fields, waste places, stream banks. Widely distributed through the United States. This species and the variety listed below furnish

good cool-season forage. Introduced from Europe. May-July.

Var. *italicum* Parn. (ANNUAL RYEGRASS). This is barely distinct from the preceding species. It tends to have larger stature, more florets per spikelet, and awned lemmas, but there are many intermediates.

Lolium temulentum L. (DARNEL). Darnel may be recognized by the very long glume, which exceeds the tip of the uppermost lemma. The lemmas are awned. Weed in grainfields and waste places, especially on the Pacific Coast and in the Southeast. This species is sometimes poisonous to livestock and human beings when eaten. Introduced from Europe.

12. Puccinellia ALKALI GRASS

The grasses of this genus have usually blunt, stiff lemmas with faint parallel nerves, and often a glistening golden or purplish band below the apex. They tend to grow in alkaline or salty places.

1a Panicle pyramidal, open, the lower branches naked near their bases; plants widespread in the United States 2

1b Panicle dense, short, the branches bearing spikelets to their bases; plants of the Atlantic Coast and desert Southwest. Fig. 110 . . . .
. . . *Puccinellia fasciculata* (Torr.) Bickn.

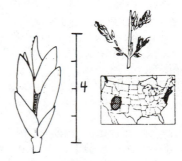

Figure 110

Perennial; tufted; plants 20-50 cm tall; panicles 5-15 cm long, stiff. Leaf blades flat, folded, or rolled, 2-4 mm wide. The species of *Puccinellia* grow on salty or alkaline wet soil. The lemmas frequently have handsome purplish, bronzy, or golden bands below the apex. This species occurs in salt marshes along the Atlantic Coast and in the arid Southwest. Introduced from Europe. June-?

2a Lower panicle branches bent downward; lemmas blunt, broadest near the apex. Fig. 111. . . . . *Puccinellia distans* (L.) Parl.

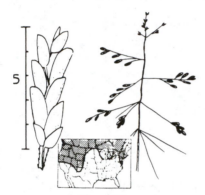

Figure 111

Perennial; tufted; culms erect or decumbent, 20-50 cm tall; panicle 5-15 cm long. The stiffish, drooping lower panicle branches are characteristic. Leaf blades flat or rolled, usually 2-4 mm wide. Moist soil. Introduced from Europe but now widespread in the United States. This species is rapidly increasing in the Midwest along highways, where it is favored by winter salt applications, and is even being planted for road shoulder stabilization. June-August.

2b Lower panicle branches not bent down; lemmas acute, broadest near the middle. Fig. 112. . . . . ALKALI GRASS *Puccinellia airoides* (Nutt.) Wats. & Coult.

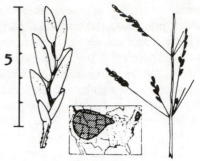

Figure 112

Perennial; tufted; plants 30-60 cm tall; panicles 10-20 cm long. Leaf blades flat or rolled, 1-3 mm wide. Similar to the preceding species, but with narrower lemmas. Native and widely distributed in the West. Sometimes cultivated under the name of Zawadke alkali grass. Also called *P. nuttalliana*. June-August.

13. Torreyochloa

The species given here are marsh and aquatic grasses, usually included in the genus *Glyceria*, from which they differ in having open sheaths and in numerous microscopic characteristics.

1a Lemmas 5 nerved; plants found west of the 100th meridian. Fig. 113 . . . . *Torreyochloa pauciflora* (Presl) Church

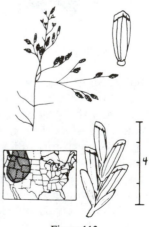

Figure 113

Perennial; tufted; leaf blades usually 10-15 cm long; 5-15 mm wide; plants 50-120 cm tall; panicles drooping, 10-20 cm long; spikelets 4-5 mm long, usually with 5-6 florets. The broad lemmas have a purple line near their translucent apex. Marshes, wet meadows, and shallow water, up to timber line. June-September.

1b Lemmas 7 nerved; plants found east of the Mississippi. Fig. 114 . . . . *Torreyochloa pallida* **(Torr.) Church**

Figure 114

Perennial; plants weak and sprawling; culms 30-100 cm long; foliage glabrous; leaf sheaths split; blades usually 4-8 mm wide; panicles open, 5-15 cm long; spikelets usually 6-7 mm long, with 4-7 florets; tips of lemmas thin, membranous, irregular; lemmas 2.5-3 mm long. This species is found in cold, wet places, often in shallow water. May-June.

Var. *fernaldii* is a late-blooming form with very narrow leaf blades, 1-3 mm wide. Wet places and shallow water; Newfoundland to Pennsylvania, westward to Minnesota and Wisconsin.

14. Scolochloa
The single species in this genus is a succulent marsh grass, often missed because its habitats are difficult to visit. The tips of the lemmas are irregularly toothed. Fig. 115. . . . . *Scolochloa festucacea* **(Willd.) Link**

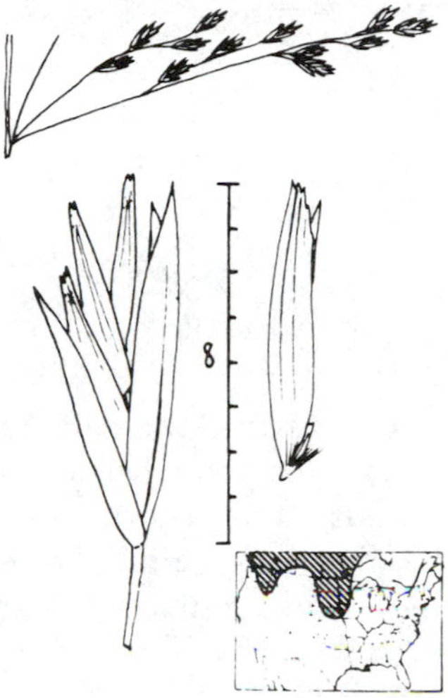

Figure 115

Perennial; spreading by thick, elongated rhizomes; plants 1-1.5 m tall; panicles open, 15-20 cm long; leaf blades usually 5-10 mm wide. This species is a plant of marshes and stream borders, where it often forms large colonies along with sedges, rushes, and other kinds of aquatic grasses. Furnishes some forage and marsh hay. Also known as *Fluminea festucacea*. Late June and early July.

15. Catapodium
The single species present in this country is a rather stiff small grass, resembling small species of *Festuca* or *Poa* and differing mostly in the short, thick, stiff pedicels and branches. Fig. 116 . . . . **FERN GRASS** *Catapodium rigidum* **(L.) C.E. Hubb.**

Figure 116

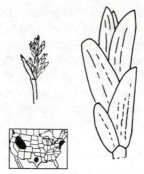

Figure 117

Tufted annual, in small clumps, 10-20 cm tall, the panicles half the total height, with stiff triangular rachis and short, thick, stiff triangular branches and pedicels; spikelets few, with short 1-3 nerved glumes; florets usually 5-10, the nerves of the lemmas inconspicuous. A weedy grass of open dry soil in the eastern, southern, and western states. It has been called *Scleropoa rigida* in U.S. manuals.

16. Sclerochloa
Diminutive annual, with short, thick, 1-sided spikes or spikelike panicles, the short, stiff branches bearing 1-2 rigid, strongly keeled spikelets; first glume 3 nerved; second glume 7 nerved; lemmas 5 nerved, both glumes and lemmas parallel-nerved, rather blunt; florets usually 3, rarely more. Disarticulation is stated to be by the fall of entire spikelets, but I have seen spikelets that disarticulate just below the third floret. Fig. 117. . . . .
. . . . *Sclerochloa dura* (**L.**) **Beauv.**

The small tufted plants, usually less than 10 cm tall, grow on disturbed dry soil. The rigid inflorescence may have a few short branches at its base, but bears only single spikelets toward the tip of the rachis. *Sclerochloa* is an insignificant weed of European origin, found in scattered localities in the intermountain and northwestern states as well as in central Texas and the northeastern states. The plants form short tufts or flat mats. May-June.

17. Poa BLUEGRASS
This is a large and important genus of annual and perennial grasses of temperate and cold climates. There is no single character which defines the genus. Many species have a "web" or coma of crimped cottony hairs attached to the callus. The spikelets are usually small, on thin pedicels. The lemmas are never awned or sharp-pointed. The 5 nerves of the lemma may be hairy or glabrous. The leaf tips are blunt and cupped or "boat-shaped." Many species are valuable as lawn, pasture, or native forage grasses.

1a **Plants producing rhizomes** 2

1b **Plants tufted, without rhizomes.** 9

2a **Culms round or nearly so in cross section (in pressed specimens, at least the nodes will be round)** 3

2b Culms strongly flattened, lens-shaped in cross section; spikelets never woolly. Fig. 118.........................
.............CANADA BLUEGRASS
Poa compressa L.

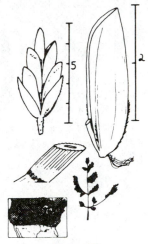

Figure 118

Perennial; rhizome-bearing; panicles narrow, 3-7 cm long. The plants are rather wiry and form looser turf than Kentucky bluegrass. It is somewhat more drought resistant than *Poa pratensis,* but yields less forage. The plants bloom several weeks later than Kentucky bluegrass in the same locality. Introduced from Europe. Mid-May-September.

3a Lemmas completely glabrous or minutely scabrous, or with a small web of cottony hairs (Fig. 119) attached to the callus ... 4

Figure 119

3b Lemmas pubescent on the keel or nerves, sometimes with a web at the base also .. 5

4a Panicles dense, with short branches, the spikelets overlapping; southern plains states or Southeast.................7b

4b Panicles open, the slender spreading branches naked at the base; western mountains......................6b

5a Lemmas pubescent only on the nerves; glabrous between them 6

5b Lemmas pubescent on the nerves and also between them near the base. Fig. 120.........................
.............PLAINS BLUEGRASS
Poa arida Vasey

Figure 120

Perennial; producing rhizomes; culms 20-60 cm tall; leaf blades borne mostly near the base of the plant, folded, 2-3 mm wide; panicle dense, cylindrical, 2-10 cm long, with short branches; spikelets 5-7 mm long, the first glume with one nerve; anthers about 1.5 mm long. Salty or alkaline meadows, up to 3000 m elevation; an important forage grass on the Great Plains. June-?

Poa glaucifolia Scribn. & Will. is similar. The first glume has 3 nerves; anthers about 2.5 mm long; panicles more open; herbage glaucous. Moist ground and open woods; Minnesota to British Columbia, to Nevada and New Mexico.

6a Lemmas bearing a web of cottony hairs at the base; sheaths glabrous 7

6b Lemmas without a web of hairs; lower sheaths glabrous or pubescent. Fig. 121. . . . . *Poa nervosa* (Hook.) Vasey

Figure 121

Perennial; producing rhizomes; culms 30-70 cm tall, in large, leafy tufts; leaf blades flat or folded; ligules 1-2 mm long; panicles open, usually 5-10 cm long, with drooping branches, naked at the base. The lemmas vary from entirely glabrous to hairy on the nerves or scaberulous. This species is highly variable in the hairiness of both sheaths and lemmas. In common with most of the native and introduced bluegrasses, it is a valuable forage plant for domestic grazing animals and wild herbivores. Dry soil in open woods, intermediate altitudes. May-August.

7a Panicles open, the long branches naked at the bases 8

7b Panicles dense and compact, cylindrical, with overlapping spikelets.

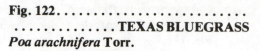

Fig. 122 . . . .
. . . . **TEXAS BLUEGRASS**
Poa arachnifera Torr.

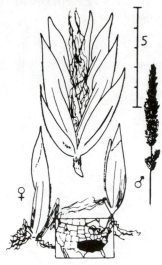

Figure 122

Perennial; rhizomes present; plants producing leafy tufts; culms 30-75 cm tall; leaf blades dark green, 2-4 mm wide, scabrous on the upper surface; panicles lobed, 5-12 cm long; spikelets usually with 5-10 florets. The staminate and pistillate spikelets are on different plants, and the two kinds are quite different in appearance. The pistillate panicles are woolly due to the presence of copious cottony webs on the lemmas. The staminate inflorescences are not woolly, their lemmas having only small webs or none at all. The plants are succulent and very leafy; they provide excellent forage in winter and spring. Sometimes cultivated for forage in the southeastern states. April-May.

Poa macrantha Vasey is also dioecious. It is a sand-binder on the coastal dunes from Washington to northern California. The plants have extensive rhizomes and stolons. Spikelets large, about 12 mm long; webs scanty.

8a Lower panicle branches mostly in pairs; lemmas 3.5-5.4 mm long; anthers 2.0-3.5

mm long. Fig. 123 . . . .
. . . . *Poa cuspidata* Nutt.

Figure 123

Perennial; rhizomes present; culms 30-50 cm tall; plants growing in large tufts, with numerous long basal leaves nearly as long as the culms; leaf blades soft, 2-3 mm wide; upper culm blades very short, rounded to an abrupt tip; panicles 7-12 cm long, very open, pyramidal, the spikelets all borne at the outer tips of the branches. This is the earliest blooming of all the native eastern grasses, usually beginning to flower in early April in the North. It is a plant of the central and southern Appalachians, where it is commonly found on rocky banks and dry wooded hillsides. Unlike Kentucky bluegrass, which it resembles, it completely lacks aggressive or weedy tendencies.

8b **Lower panicle branches mostly in whorls of 5; lemmas 2.4-3.6 mm long; anthers 1.0-1.8 mm long. Fig. 124** . . . .
. . . . **KENTUCKY BLUEGRASS**
Poa pratensis L.

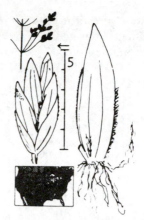

Figure 124

Perennial; rhizomes present; culms 30-100 cm tall, in dense clumps with numerous sterile leafy shoots (innovations); panicles open, pyramidal, somewhat contracted after flowering. Kentucky bluegrass is one of our most widely distributed introduced grasses and is much used for lawns and pastures in the northern states. It is extensively naturalized in pastures, prairies, roadsides, open woods and waste ground. Introduced from Europe. April-July.

9a **Lemmas bearing a web** **10**

9b **Lemmas glabrous or pubescent, but without a web** **15**

10a **Marginal nerves of lemmas pubescent** **11**

10b **Marginal nerves of lemmas glabrous** . . **14**

11a **Lemmas glabrous between the nerves; ligules of upper culm leaves 2.5-7 mm long** **12**

11b **Lemmas pubescent between the nerves; ligules almost always less than 2 mm long** **13**

12a Lemmas greenish-yellow, thin, with prominent intermediate nerves; sheaths usually scaberulous. Fig. 125 . . . .
. . . . . . **ROUGH STALK BLUEGRASS;**
TRIVIALIS *Poa trivialis* **L.**

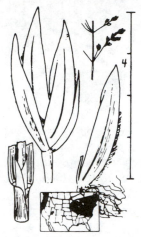

Figure 125

Perennial; tufted; culms 30-100 cm tall, often somewhat bent or reclining at the base; panicles open, ovoid or pyramidal, the spikelets clustered near the outer ends of the branches; lemmas glabrous or hairy on the lateral nerves. The plants are often found growing wild in wet areas, and the seed is planted in lawn mixtures for shady places. Introduced from Europe. May-June.

12b Lemmas green, usually purple or bronzy at the tip, the intermediate nerves inconspicuous; sheaths smooth. Fig. 126 . . . .
. . . . *Poa palustris* **L.**

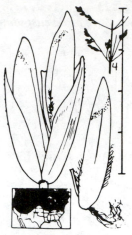

Figure 126

Perennial; plants slender, weak, with bent and reclining (decumbent) culm bases; panicles open, pyramidal, with slender, weak branches. The lemmas are usually purple and golden banded near the tip, making them very attractive under the lens. Wet meadows, stream banks, and moist woods; seldom numerous in one place. This species is native to both North America and Eurasia. June-August.

Poa interior Rydb. is very similar to the above, but has erect, tufted culms, small panicles less than 10 cm long, and short ligules usually 1 mm long or shorter. Across Canada and south to northern New England, Wisconsin, North Dakota and the High Plains and western mountain states.

13a Panicles cylindrical, 10-20 cm long, the branches 3-6 in a whorl; lower branches drooping . . . .
. . . . See *P. sylvestris* under *P. alsodes*, Fig. 127.

13b Panicles pyramidal, 5-10 cm long, the branches 1-2 at a node, ascending . . . . . .
. . . . See *P. interior* under *P. palustris*. Fig. 126.

14a Upper ligules 3-7 mm long; sheaths usually scaberulous; intermediate nerves of lemmas conspicuous **12a**

14b Upper ligules 0.5-2 mm long; sheaths smooth; intermediate nerves of lemmas inconspicuous. Fig. 127 . . . .
. . . . *Poa alsodes* A. Gray

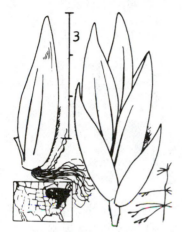

Figure 127

Perennial; tufted; plants slender, 30-60 cm tall, erect and graceful, with very open, delicate, cylindrical panicles, 10-25 cm long and half as wide. The slender branches bear a few spikelets near their tips. Cool rocky woods or wooded flood plains in the northeastern states and the Appalachians. May-June.

Poa sylvestris A. Gray is found also in rich woods and resembles this species, but has lemmas which are hairy between the nerves. New York to Wisconsin, southward to Florida and Texas. May-July.

15a Lemmas pubescent between the nerves, the hairs often very short **16**

15b Lemmas glabrous between the nerves. . **17**

16a Panicles very open, the branches slender, spreading, usually in pairs, bearing spikelets only near their outer ends; lemmas with conspicuous long hairs on their keels; forested eastern half of the United States, to Michigan and Texas. Fig. 128 . . . .
. . . . *Poa autumnalis* Muhl.

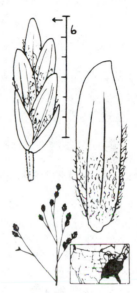

Figure 128

Perennial; tufted; plants delicate, 30-60 cm tall; panicles 10-20 cm long and nearly as wide, their branches very slender, bearing a few spikelets near the tips. Leaf blades 2-3 mm wide, many at the base of the plants. The plants are not, as the name would seem to indicate, autumn blooming. Moist woodlands. June.

16b Panicles usually dense, the branches bearing spikelets nearly to their bases; nerves of lemmas not bearing long hairs; western states, eastward to Minnesota and to the Dakotas. Fig. 129 . . . .
. . . . *Poa scabrella* (Thurb.) Benth.

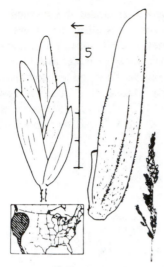

Figure 129

Perennial; tufted; culms 50-100 cm tall; panicles slender, elongated, with ascending branches. This species and its relatives have lemmas covered, at least near their bases, with short crimped or appressed hairs. A number of very closely related species or forms exist in this group, differing mostly in the shape of the panicle. All are bunch grasses of the western states and are valuable forage grasses at lower and medium altitudes. March-August.

17a **Midnerve and lateral nerves of lemmas pubescent** **18**

17b **Lemmas glabrous or minutely scaberulous, not pubescent** **19**

18a **Spikelets 6-8 mm long; keels and marginal nerves of lemmas pubescent; intermediate nerves inconspicuous, not pubescent. Fig. 130** . . . . **MUTTON GRASS** *Poa fendleriana* **(Steud.) Vasey**

Figure 130

Perennial; erect, densely tufted bunch grass; leaves mostly basal, stiff and scabrous, folded or rolled, 1-2 mm wide; culms 30-50 cm tall, not much exceeding the basal leaves; panicles dense, oblong, 2-7 cm long, the pale spikelets overlapping. The plants are dioecious but the spikelets of both sexes are similar. Ligules short, less than 1 mm long. Medium altitude hills and dry forests in the western mountains, from 2300-4000 m. It is regarded as one of the best of western forage grasses, especially for sheep. April-July.

Poa longiligula Scribn. & Will. is very similar, but has ligules up to 5-7 mm long. Range about the same as the preceding species.

18b **Spikelets 4-6 mm long; keels, intermediate nerves, and marginal nerves of lemmas conspicuous, usually all pubescent. Fig. 131** . . . .
. . . . **ANNUAL BLUEGRASS** *Poa annua* **L.**

Figure 131

19b Panicles elongated, cylindrical, 10-15 cm long; spikelets scarcely keeled, narrowly elliptical in outline, 3 or more times longer than wide, the glumes and lemmas rounded on the back; nerves of lemmas inconspicuous. Fig. 132 . . . . *Poa nevadensis* Vasey

Figure 132

Annual; tufted. The diminutive light green plants, usually less than 20 cm tall, are soft, weak, and spreading. Panicles 3-7 cm long, pyramidal, with short, spreading branches. The plants sometimes grow in shallow water, then becoming long, slender, and rooting at the nodes. This species begins growth in fall or early spring and blooms very early, and may die out when hot weather comes. Some blooming also occurs in fall. Lawns, paths, margins of water and open woods, roadsides. Introduced from Europe.

Poa chapmaniana Scribn., a native species, is very similar, but the lemmas have a small web at the base; anthers very small, 0.2 mm long. Southern New England to Nebraska and southward.

Perennial; tufted; culms erect, 50-100 cm tall; sheaths scabrous; panicle narrow, elongated, 10-15 cm long. This is one of a group of very similar bunch grasses of the western mountain states. All have slender, elongated spikelets, whose lemmas are not keeled; lemmas glabrous and with inconspicuous nerves. They furnish excellent range forage for wild game animals and domestic livestock. Plains, dry meadows; open or partially wooded mountainsides, from near sea level up to 3700 m elevation. May-September.

19a Panicles short, ovoid, 2-8 cm long; spikelets ovate in outline, about twice as long as wide; glumes and lemmas keeled; nerves of lemmas rather conspicuous . . 20

20a Leaf blades scabrous; ligules of culm leaves less than 2 mm long. Fig. 133 . . . . *Poa cusickii* Vasey

Figure 133

Perennial; culms in dense bunches, 20-60 cm tall; basal leaf blades thread-like, very scabrous, about half the length of the culms; ligules usually under 1 mm long; panicles 3-8 cm long, dense, oblong, pale or somewhat purplish. Rocky slopes and sagebrush plains, medium to high altitudes. May-July.

20b Leaf blades glabrous; ligules of culm leaves 2-4 mm long. Fig. 134 . . . .
. . . . *Poa epilis* Scribn.

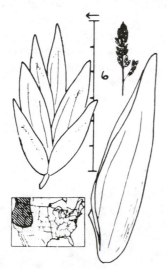

Figure 134

Perennial; culms in small tufts; plants 20-40 cm tall; leaf blades of the culms short, flat, 2-3 mm wide, their ligules 2-4 mm long; basal leaves long, folded or rolled; panicles on long, thin peduncles; panicles short, dense, oblong, 2-6 cm long; spikelets about 5 mm long, usually with 3 florets; lemmas 4-6 mm long, sometimes scaberulous. *Poa epilis,* sometimes called "skyline bluegrass," is an important high altitude forage grass, usually found above timber line on steep slopes and in mountain meadows. July-August.

About five similar species are to be found at high altitudes in the western mountains, mostly above timber line. They are all dwarf, rarely more than 20 cm tall.

18. Briza QUAKING GRASS
Delicate annual grasses, with spikelets on slender drooping pedicels; florets of the spikelets placed at right angles to the rachilla; lemmas nearly circular, without visible nerves. Fig. 135 . . . .
. . . . **LITTLE QUAKING GRASS**
Briza minor L.

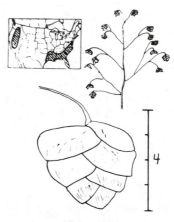

Figure 135

Annual; tufted; 10-40 cm tall. The species of *Briza* have very slender pedicels, which allow the drooping spikelets to quiver in any breeze. Ligules 4-5 mm long. This species is found as a weed in the southern states and often in California. Introduced from Europe. April-May.

Briza media L. is similar, but has larger spikelets. The ligules are about 1.5 mm long. Northeastern states.

Briza maxima L., with spikelets 10 mm wide, is the largest species. It is sometimes grown for winter bouquets.

19. Catabrosa
Spikelets 1-2 flowered, densely clustered along the panicle branches; lemmas with wide blunt tips and 3 parallel nerves; disarticulation above the glumes and between the florets. Fig. 136 . . . .
. . . . **BROOKGRASS** Catabrosa aquatica **(L.) Beauv.**

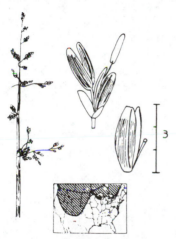

Figure 136

Perennial. The culms may lie on the ground and root at the nodes for half or more of their length of 10-50 cm. Panicles yellowish, 10-20 cm long, open, pyramidal. Brookgrass is a soft, succulent grass of wet ground, found in the subarctic and at higher elevations in the western mountains of the United States. The plants make excellent summer feed for livestock. Found also in northern Europe and Asia. June-August.

20. Dactylis
Panicles with a few stiff, rigid branches; spikelets nearly sessile in dense, 1-sided tufts at the ends of the branches; glumes and lemmas ciliate on the keels; lemmas pointed or short-awned. Fig. 137 . . . .
. . . . **ORCHARD GRASS**
Dactylis glomerata **L.**

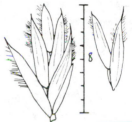

Figure 137

Perennial; tufted; plants 60-120 cm tall; herbage light green; leaf blades 2-8 mm wide. The panicle branches spread only at flowering time. The plants grow in large tussocks. They furnish both pasturage and hay, and this species is rather important as a forage grass. It will tolerate partial shade. Introduced from Europe. May-July.

21. Cynosurus
Inflorescence a dense, somewhat 1-sided panicle with paired fertile and sterile spikelets; sterile spikelets awnless, not disarticulating; fertile spikelets awned, disarticulating into single florets. Fig. 138 . . . .
. . . . **CRESTED DOGTAIL**
Cynosurus cristatus **L.**

Figure 138

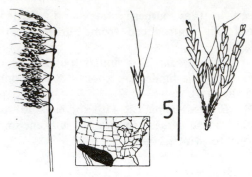

Figure 139

Perennial; tufted, culms 30-60 cm tall. The sterile spikelets are paired with fertile ones in the same inflorescence. Sterile spikelets are made up of slender empty awned lemmas. Inflorescence a dense spikelike panicle. Dogtail is a European grass which has been imported for use in lawn and meadow seed mixtures. It is occasionally found in lawns, pastures and waste places in the eastern states and also in the Pacific Northwest. Apparently it has little value for forage. June-August.

Cynosurus echinatus L. has a short, bristly, 1-sided panicle. The lemmas have awns up to 10 mm long. East Coast; Pacific Northwest; south central states.

22. Lamarckia

Small annual grasses; panicle dense and fluffy, composed of drooping yellow or purplish spikelet clusters; the spikelets of each cluster falling as a unit of 1 perfect and several sterile spikelets; fertile spikelet with a single long-awned perfect floret and a reduced rudimentary one; sterile spikelets composed of numerous awnless empty florets. Fig. 139. . . . .
. . . . **GOLDENTOP**
Lamarckia aurea (L.) **Moench**

Goldentop occurs as a weed and may be cultivated in the southwestern states. February-June.

Tribe 5. Aveneae

23. Avena OATS

Tufted annuals; panicles and spikelets large; florets usually with stout twisted awns borne on the back of the lemma; florets usually 2 or 3, rather rigid; grain usually retained in the floret; disarticulation above the glumes and between the florets; callus often bearded. Fig. 140. . . . . .
. . . . **WILD OATS** *Avena fatua* L.

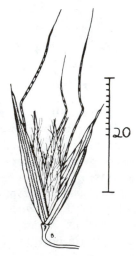

Figure 140

Annual; culms usually 30-100 cm tall, in small tufts; panicles open, up to 30 cm long. The spikelets when ripe are open, bell-shaped, with conspicuous protruding bent awns. The scar at the base of the floret is circular and prominent and is usually called a "sucker mouth." Wild oats is distinguished by the hairy lemmas, prominent awns and the sucker mouth. It is widely dispersed in the United States but is most common in the Pacific coast states, where it is a prevalent weed and is sometimes cut for hay. May-August.

Avena sativa L. **(CULTIVATED OATS)** Fig. 141, differs from wild oats in having lemmas which are glabrous except on the callus, no sucker mouth, and weak, usually straight awns. Widely cultivated and growing from scattered seed on roadsides and waste ground.

Figure 141

24. Helictotrichon
Tufted perennials; spikelets with 2-6 florets; lemmas stiff, awned from the back; rachilla hairy; spikelets erect. Fig. 142 . . . .
. . . . **SPIKE OAT**
Helictotrichon hookeri **(Scribn.) Henr.**

Figure 142

The species of this genus are sometimes placed in the genus *Avena,* but differ in the characters given above. Plants of this species are found on grasslands and dry mountain slopes in the Rocky Mountains and the northern plains states. July-August.

25. Arrhenatherum
Perennial, often with bulbous bases; spikelets with 2 florets, the lower one staminate, with a bent, protruding awn; upper floret perfect, with a short straight awn; spikelets 7-9 mm long; rachilla prolonged beyond the base of the second floret. Fig. 143 . . . .
. . . . **TALL OATGRASS**
Arrhenatherum elatius **(L.) Mert. & Koch**

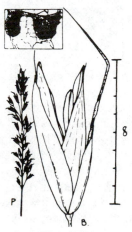

Figure 143

Perennial; tufted; culms 1-1.5 m tall; leaf blades 5-10 mm wide; panicles 15-30 cm long; narrow and elongated, but with spreading short branches; glumes thin and somewhat translucent. Rarely both florets have bent awns. Tall oatgrass is cultivated in the northern states as a meadow grass and has freely escaped to roadsides and waste ground. Introduced from Europe. May-September.

26. Deschampsia
Tufted, usually perennial grasses; spikelets small, with 2 florets; disarticulation above the glumes and between the 2 florets; rachilla prolonged beyond the second floret as a hairy bristle; awn attached below the middle of the lemma, which has several teeth at the apex. Fig. 144 . . . .
. . . . **HAIRGRASS**
Deschampsia caespitosa **(L.) Beauv.**

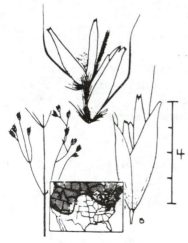

Figure 144

Perennial; tufted; culms slender, 60-120 cm tall; leaves flat and scabrous, mostly at the base of the plants, 1-4 mm wide; panicles open, delicate; spikelets often purple. The awns are nearly hidden within the glumes. The lemmas have several minute teeth at their tips. Bogs, wet ground, mountain meadows. An important forage grass in the West. May-July.

Deschampsia flexuosa (L.) Trin. is similar but the awns are strongly bent and protrude from the spikelets. Leaf blades fine and hairlike. Arctic North America, southward to Minnesota and Michigan and southward in the Appalachian Mountains to Georgia; Arkansas and Oklahoma.

27. Aira
Delicate annual grasses; spikelets small, 2 flowered, the lemmas with 2 teeth at the tip; awn attached to the back of the lemmas below the middle; rachilla not prolonged beyond the base of the upper floret. Fig. 145 . . . .
. . . . **SILVER HAIRGRASS**
Aira caryophyllea **L.**

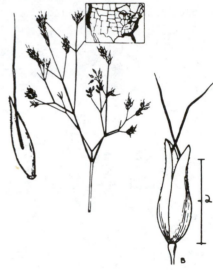

Figure 145

Annual; tufted; culms 10-35 cm tall. The plants look like delicate little trees with open crowns. The silvery spikelets are closely clustered at the tips of the branches. Dry ground, mostly in the Atlantic and Pacific coastal regions. Introduced from Europe. April-July.

Aira elegans Willd. is similar but the lower floret is awnless or nearly so. Atlantic and Gulf Coastal Plains; Oregon and California. May-June.

28. Trisetum

Tufted, usually perennial grasses; spikelets small, 2 flowered; rachilla hairy, extended beyond the base of the second floret as a hairy bristle; lemma 2 toothed at the tip; awn, if present, attached above the middle of the lemma.

1a **Awn short, concealed within the glumes, or absent. Fig. 146.** . . . .
. . . . *Trisetum wolfii* **Vasey**

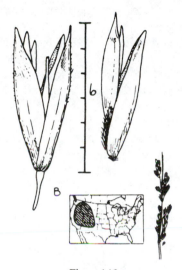

Figure 146

Perennial; tufted or sometimes with short rhizomes; culms 50-100 cm tall; leaf blades flat, 2-4 mm wide, scabrous on the upper surface; panicles dense, cylindrical, yellowish. An important forage grass in moist mountain meadows at mid-altitudes in the western mountains. July-September.

Trisetum melicoides (Michx.) Scribn. is similar but has a loose, drooping panicle. Moist shores and swamps. Newfoundland and New England to Wisconsin. August.

1b **Awn protruding from the glumes, bent**
and twisted. Fig. 147 . . . .
. . . . **SPIKE TRISETUM**
Trisetum spicatum **(L.) Richt.**

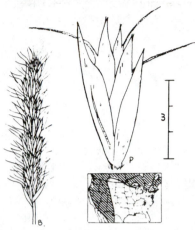

Figure 147

Perennial; tufted; culms 15-50 cm tall; foliage smooth or finely hairy; panicles spikelike, dense, shaggy with many protruding awns, purplish or golden. An important forage grass at high altitudes, on slopes and in mountain meadows in the western mountains. The plants furnish good forage throughout the growing season, especially for cattle and horses. June-August.

29. Sphenopholis

Annual or perennial tufted grasses; inflorescence usually dense or spikelike; spikelets disarticulating below the glumes, sometimes with a stub of the pedicel attached; first glume narrow, the second usually much broader; florets 2, awnless or awned; rachilla prolonged beyond the second lemma; spikelets very flat.

1a **Second floret with a bent awn arising between 2 teeth. Fig. 148** . . . .
. . . . *Sphenopholis pennsylvanica* **(L.) Hitchc.**

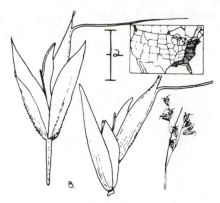

Figure 148

Perennial; culms slender and weak, 50-100 cm tall. The spikelets fall with about half of the pedicel attached. Meadows, swamps, and wet ground. May-June. Formerly called *Trisetum pennsylvanicum*. Recent studies have shown that this species crosses freely with species of *Sphenopholis,* producing sterile, awned hybrids which have been called *Sphenopholis pallens.*

Trisetum interruptum Buckl. has similar spikelets but a dense spikelike panicle. Open dry plains, Texas to Colorado and Arizona. March-May.

1b Both florets lacking awns; lemmas not toothed at the tip. Fig. 149 . . . . WEDGEGRASS
Sphenopholis obtusata (Michx.) Scribn.

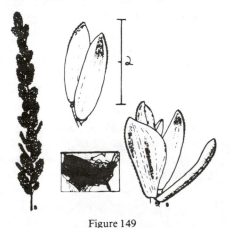

Figure 149

Perennial; tufted; culms 30-100 cm tall; panicles dense and cylindrical, the branches spreading somewhat at flowering time. The peculiarly shaped glumes are the most characteristic thing about the wedgegrasses. In this species, the second glume is 2-2.5 times longer than its folded width. Succulent and probably good feed, but not occurring in dense stands. May-August.

Var. *major* (Torr.) Erdman has a looser, more open panicle; the second glume is 3-4 times longer than the folded width. Throughout the United States and southern Canada. May-July.

Sphenopholis nitida (Biehler) Scribn. has a slender, open panicle. The upper lemma of each spikelet is visibly scabrous under a lens. Woods, eastern states to Illinois and Texas.

30. Koeleria
Spikelets flattened; glumes nearly equal, shorter than the florets, the second wider than the first; lemmas awnless or with a very short awn arising at the split tip of the lemma; rachilla joints very short; disarticulation above the glumes and between the florets. Fig. 150 . . . . **JUNEGRASS**
Koeleria cristata (L.) Pers.

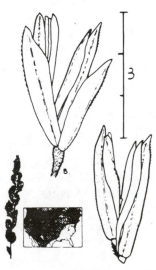

Figure 150

Perennial; tufted; culms 30-60 cm tall; panicles yellowish or silvery in color, narrowly cylindrical or somewhat lobed. At blooming time the branches spread but later close up again. Spikelets with 2-4 florets. Junegrass is one of the most widely distributed of American grasses. Dry or sandy soil; prairies or open woods. Also found in Eurasia. June-September.

31. Holcus

Spikelets 2 flowered, flat, disarticulating below the glumes; foliage soft, velvety. Fig. 151 . . . . **VELVET GRASS** *Holcus lanatus* **L.**

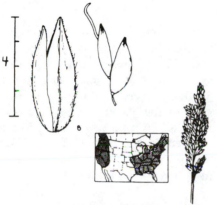

Figure 151

Perennial; tufted; culms 30-60 cm tall; panicles elliptical, closely flowered. The spikelets have an awnless perfect lower floret and a staminate second floret with a short awn. The entire plant is velvety to the touch. Open moist ground, meadows, thickets. Velvet grass was imported for forage but is rarely cultivated now. Forage value low. Introduced from Europe. June-August.

32. Beckmannia

Inflorescence a panicle of short, 1-sided spikes, the circular spikelets in 2 rows along the lower side of the rachis of each spike; spikelets disarticulating below the glumes; floret 1 (2 in the

European species). This genus was formerly placed with the Chlorideae because of its inflorescence. Its chromosomes and microscopic characters indicate its proper position is in the Pooideae. Fig. 152 . . . . **SLOUGH GRASS** *Beckmannia syzigachne* **(Steud.) Fern.**

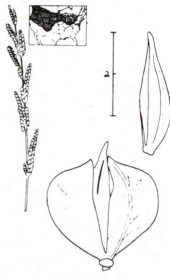

Figure 152

Annual; tufted; plants up to 100 cm tall. The slender, wand-like panicles of spikes produce large quantities of wrinkled, circular spikelets which shatter from the plants at a touch. Plants rather tender and succulent; forage value good. Wet meadows, shores of ditches and lakes, mostly in the western states. The plants are sometimes cut for hay. July-August.

Tribe 6. Phalarideae

33. Anthoxanthum

Glumes longer than florets, the second about twice as long as the first; florets 3, the lower 2 sterile and awned, the uppermost fertile and awnless, all 3 falling from the glumes as a unit. Fig. 153 . . . . **SWEET VERNAL GRASS** *Anthoxanthum odoratum* **L.**

Figure 153

Figure 154

Perennial; tufted; plants 30-60 cm tall; panicles yellowish-brown, cylindrical, dense, usually less than 6 cm long; glumes very thin and delicate. The 2 hairy sterile lemmas are split at the tip and awned from the middle of the back. Sweet vernal grass has the delightful fragrance of coumarin, which persists long after the plants are dried. This species was introduced from Europe, presumably as a meadow plant to add fragrance to hay. Now it is widely established as a weed of roadsides, woods, and meadows, except in the dry parts of the United States. One of the earliest of spring grasses, blooming from the middle of April to July in the northern states.

34. Hierochloë
Glumes equal, very thin, longer than the florets; florets 3, the lower 2 usually awnless, staminate; the uppermost floret perfect-flowered, stiff; panicles brownish, shining. Fig. 154 . . . .
. . . . **HOLY GRASS**
Hierochloë odorata (L.) **Beauv.**

Perennial; culms 30-60 cm tall, single or in small tufts, arising from slender creeping rhizomes. Holy grass is an attractive species, with handsome shining golden-brown panicles. The glumes are very thin and translucent. The plants have the sweet, vanilla-like scent of coumarin, and were used by the American Indians as material for basketry, and in Europe as perfume in certain religious ceremonies. Moist meadows, bogs, and prairies. April-July.

35. Phalaris CANARY GRASS
Inflorescence a dense, cylindrical or thimble-shaped panicle; spikelets strongly compressed, the glumes keeled and usually winged, equal and longer than the concealed florets; fertile terminal floret rigid, shining; falling from the glumes with 2 minute, scale-like sterile florets attached at the base.

1a Plants producing rhizomes; panicles cylindrical and often lobed. Fig. 155 . . . .

. . . . REED CANARY GRASS
Phalaris arundinacea L.

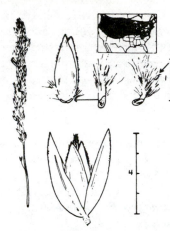

Figure 155

Perennial; plants 60-150 cm tall, with pale or purplish panicles, 7-16 cm long. The tall, broadleaved plants furnish considerable forage and are often cut for hay. They form dense colonies in marshes and along ditches. Reed canary grass is now planted for erosion control on farm waterways. Forms with white-striped leaves are sometimes grown for ornament, under the name of "gardeners garters." May-August.

1b Plants tufted, lacking rhizomes; panicle dense, thimble-shaped. Fig. 156. . . . .
. . . . CANARY GRASS
Phalaris canariensis L.

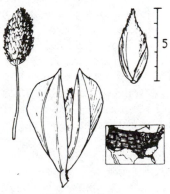

Figure 156

Annual; tufted; culms 30-60 cm tall; panicles 1-4 cm long. The strongly winged glumes are striped with green and white lines. This is the species which furnishes the "canary seed" which is fed to caged birds. Occasionally the plants are found growing on trash heaps where the sweepings from bird cages are deposited, but this species apparently does not grow as a truly wild plant in this country. Introduced from Europe. June-August.

A number of other similar species are found in various parts of the United States. None is common or conspicuous.

Tribe 7. Agrostideae

36. Ammophila
Glumes longer than the floret; lemma awnless, bearded on the callus; rachilla prolonged behind the palea as a hairy bristle. Fig. 157 . . . .
. . . . AMERICAN BEACHGRASS
Ammophila breviligulata Fernald

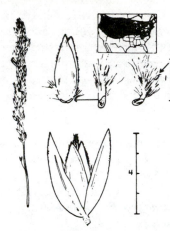

Figure 157

Perennial; spreading by very long, stiff rhizomes. The plants are coarse and tough. Panicles dense and cylindrical, 10-20 cm long. The plants are highly important sand-binders on dunes along the Atlantic Coast and on sand beaches of the Great Lakes. July-September.

A similar species with long ligules, 10-

30 mm long, *Ammophila arenaria,* has been planted on the sand dunes along the Pacific Coast. European.

37. Calamagrostis REEDGRASS

Perennials; inflorescence an open or dense panicle; spikelets small, glumes equal, longer than the floret; lemma awned from below the middle, with a tuft of long straight hairs on the callus; rachilla extended behind the palea as a thin, often hairy bristle. This may be concealed by the callus hairs and must be sought carefully.

1a   **Awn straight, hidden within the glumes; hairs nearly as long as the lemma 2**

1b   **Awn bent sidewise, protruding from the glumes; hairs shorter than the lemma. Fig. 158 . . . .**
    . . . . PINEGRASS
Calamagrostis rubescens **Buckl.**

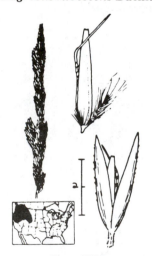

Figure 158

Perennial; culms in tufts, 60-100 cm tall; plants producing rhizomes; leaf blades scabrous, 2-4 mm wide, flat or somewhat rolled; panicles dense and cylindrical, 7-15 cm long, pale or purplish in color; glumes 4-5 mm long; sterile rachilla joint about 1 mm long, its hairs about twice as long. While common, pinegrass is low in palatability, especially for sheep, and is little grazed except when young and green. The plants make a strong, tough turf which resists heavy grazing and trampling. Most of the reproduction is by rhizomes. Coniferous forests, up to 3300 m elevation. June-August.

2a   **Panicle open, pyramidal, with spreading or drooping elongated branches. Fig. 159 . . . .**
    . . . . BLUEJOINT
Calamagrostis canadensis **(Michx.) Beauv.**

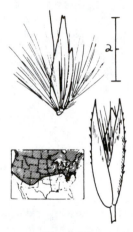

Figure 159

Perennial; culms slender, in small tufts, 60-150 cm tall; plants producing many long creeping rhizomes. The panicles vary from rather dense to loose, pyramidal, and nodding. Bluejoint is a very widespread and common species found in marshes and wet ground. While rather tough, it furnishes a good deal of forage and is sometimes cut for marsh hay in the North Central States. June-August.

2b   **Panicle dense, cylindrical, with short, erect branches. Fig. 160 . . . .**
    . . . . *Calamagrostis inexpansa* **Gray**

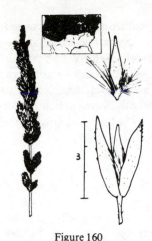

Figure 160

Perennial; culms in tufts, 40-120 cm tall; plants producing rhizomes. The dense, cylindrical panicle may be pale or purplish. The leaf blades are rough to the touch and usually rolled; ligules 4-7 mm long. Marshes and wet meadows, apparently sometimes on dry soil in the western mountains. June-July.

38. Agrostis BENTGRASS

Inflorescence an open or dense panicle; spikelets small; glumes equal, longer than and concealing the floret; lemma usually awnless, thin and delicate; palea short or obsolete, thin; rachilla not prolonged. Many species, some cultivated for lawns, pastures, and putting greens.

1a Palea at least half as long as the lemma; plants often with rhizomes or stolons. . . 2

1b Palea absent or less than one-fourth as long as the lemma; plants tufted, lacking rhizomes or stolons 4

2a Plants with erect stems; panicles open, pyramidal 3

2b Plants with creeping stolons, rooting at the nodes and forming flat mats; rhizomes absent; panicle narrow, the branches closing up after the flowering period. Fig. 161 . . . .
. . . . CREEPING BENT *Agrostis stolonifera* L., var. *palustris* (Huds.) Farw.

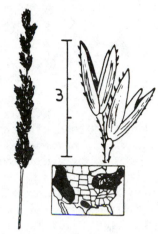

Figure 161

Perennial; low, spreading by numerous fine leafy stolons. This species is much used for golf greens and fine lawns, but requires frequent cutting and watering to produce a good turf, and the plants are subject to several serious diseases. New lawns are usually produced by planting pieces of chopped turf, which take root and spread to form a continuous sod. Reproduction by seed is also possible, but seed of many strains is very scarce or unobtainable. Various forms of this species are known as seaside bent, Coos Bay bent, Washington bent and Metropolitan bent. Marshes, especially along seacoasts, wet ground around streams, springs, lakes, and ditches. Also known from Europe and Asia. June-September.

3a Ligules 1-2 mm long; leaf blades 1-3 mm wide. Fig. 162 . . . .
. . . . BROWNTOP *Agrostis tenuis* Sibth.

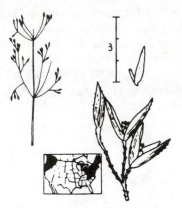

Figure 162

Perennial; plants 20-40 cm tall; rhizomes absent but short stolons sometimes present; leaf blades very narrow; ligules on sterile shoots about 1 mm long, on the culms up to 2 mm long; panicles usually 5-10 cm long, brownish, open and delicate, the spikelets all near the outer tips of the branches. *Agrostis tenuis* is cultivated as a lawn and pasture grass and is sometimes referred to as Rhode Island bent, Prince Edward Island bent, Colonial bent, New Zealand bent, or Astoria bent. The plants are occasionally found growing wild in regions where browntop is cultivated. Some forms possess lemmas which bear a delicate awn attached near the base. Introduced from Europe. June-July.

3b Ligules 3-7 mm long; leaf blades 2-6 mm wide. Fig. 163 . . . .
. . . . . REDTOP
Agrostis gigantea **Roth**

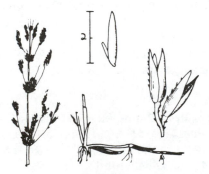

Figure 163

Perennial; culms up to 1-1.5 m tall; numerous creeping rhizomes present; panicle pyramidal, with rather dense whorls of branches, flowering to their bases. Redtop is one of our most important meadow grasses, and is also used in lawn seed mixtures. Roadsides, meadows, waste ground; very widely naturalized. Introduced from Europe. June-August.

This species has usually been called *A. alba* in U.S. manuals.

4a Panicle narrow, with the short branches bearing spikelets nearly to their bases. . . 5

4b Panicle open, the spreading branches bearing spikelets at their outer ends only 6

5a Slender alpine plants; leaf blades 5 cm long or shorter, 1-2 mm wide; ligules 1-2 mm long; panicles usually less than 5 mm wide. Fig. 164 . . . .
. . . . *Agrostis variabilis* **Rydb.**

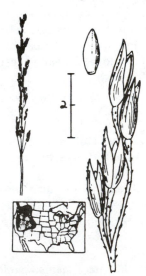

Figure 164

Perennial; tufted; culms 10-20 cm tall; panicles small, 2-6 cm long; spikelets about 2.5 mm long; lemmas about 1.5 mm long; palea very short. High altitudes in the mountains, usually above timber line along creeks and on slopes. Forage value good. July-August.

Agrostis humilis Vasey. Similar but smaller, differing from the above chiefly in having a palea about 2/3 as long as the lemma; plants 5-15 cm tall; panicles purple, slender, 1-3 cm long. Excellent forage. Bogs and mountain meadows above timber line. Montana and Washington to Colorado and Nevada. July-August.

5b **Stouter plants of medium and low altitudes; longer leaf blades 8-10 cm long; ligules 3-6 mm long; panicles usually 1-several cm wide. Fig. 165 . . . .**
. . . . SPIKE REDTOP
Agrostis exarata **Trin.**

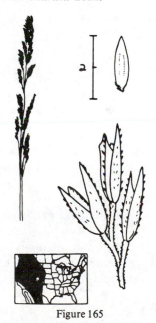

Figure 165

Perennial; tufted; culms 20-120 cm tall; panicles spikelike, either thin, or thicker and somewhat lobed. Lemma sometimes awned. This is one of the most important western range grasses. It is palatable to both domestic livestock and wild grazing animals. Usually on moist ground, in meadows, along streams, and in open woodlands. July-August.

6a **Panicles with long, slender main branches which branch again only on the outer half. Fig. 166 . . . .**
. . . . TICKLEGRASS
Agrostis scabra **Willd.**

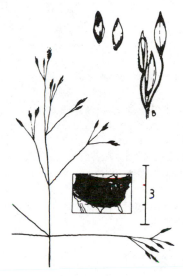

Figure 166

Perennial; culms slender, 20-80 cm tall; leaf blades threadlike, in dense tufts at the base of the culms; panicles delicate, readily breaking away from the plants and blown by the wind as tumbleweeds. Forage value fair. Ticklegrass is a very widespread species, on open ground or in partial shade, especially on moist soil. June-September.

Agrostis hyemalis (Walt.) B. S. P. is similar but has spikelets bunched at the tips of the branches. Atlantic Coastal Plain and lower Mississippi Valley. It blooms earlier, in May and early June.

6b **Main panicle branches branching again below the middle. Fig. 167 . . . .**
. . . . *Agrostis perennans*
(Walt.) Tuck.

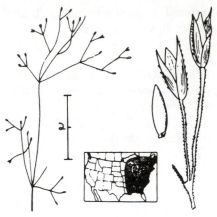

Figure 167

Perennials; tufted; culms 20-80 cm tall; panicles usually delicate and lace-like, especially in shade-grown plants. Lemmas rarely awned. Dry open fields and woodlands. July-October.

Agrostis oregonensis Vasey is similar, but larger and more vigorous. Marshes; Montana to British Columbia, southward to Wyoming and California.

39. Gastridium
Panicle dense and spikelike, bristly; glumes swollen at the base, with long, beak-like tips; rachilla extended behind the palea as a minute bristle. Fig. 168 . . . .
. . . . **NIT GRASS**
Gastridium ventricosum (Gouan) **S. & T.**

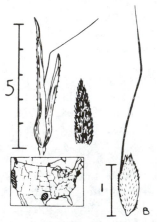

Figure 168

Annual; tufted; shallow-rooted; culms 20-55 cm tall; panicles dense, cigar-shaped, with glossy silvery or yellowish spikelets. The minute floret is concealed by the glumes. Weed of fields and waste places, common on the Pacific Coast. Introduced from Europe. May-July.

40. Polypogon
Panicles dense, soft, silky; spikelets small, with long-awned glumes and lemma; disarticulation below the spikelets, which fall with a short bit of the pedicel attached. Fig. 169 . . . .
. . . . **RABBITFOOT GRASS**
Polypogon monspeliensis (L.) **Desf.**

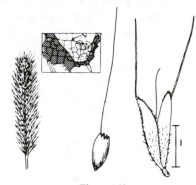

Figure 169

Annual; tufted; usually 15-50 cm tall; panicles dense, often somewhat lobed, densely covered with soft, silky yellowish awns. Rabbitfoot grass is rather widely distributed in the United States, but is most common in the West, at low altitudes. Frequently found on seepy wet ground around springs or on banks of streams. Weedy; forage value low. Introduced from Europe. May-October.

41. Scribneria
Inflorescence a slender spike or reduced panicle; spikelets 1 per node, pressed against the rachis; glumes equal, concealing the floret; floret only 1/4 to 1/2 as long as the glumes; rachilla prolonged behind the palea; awn protruding from the glumes; stamen 1. Fig. 170 . . . .
. . . . *Scribneria bolanderi* (Thurb.) **Hack.**

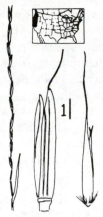

Figure 170

Slender annual grass; among shrubs; plants up to 30 cm tall, rare or inconspicuous. Lower elevations in the mountains, Washington to California. Spring.

This odd little grass was placed in the barley tribe by Hitchcock. The occasional presence of branches in the inflorescence suggests that it should not be placed in this tribe, but its microscopic characters place it in the Pooid Subfamily.

42. Limnodea
Spikelets in slender panicles, 1 flowered, disarticulating below the glumes; glumes equal, longer than the floret; lemma bearing a slender protruding awn attached below the split tip; rachilla prolonged behind the palea. Fig. 171 . .
. . . . *Limnodea arkansana* (Nutt.) Dewey

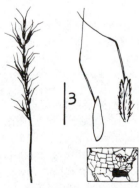

Figure 171

Tufted annual; plants 20-40 cm tall. Dry open soil; spring.

43. Cinna
Inflorescence a drooping panicle; spikelets disarticulating below the glumes; lemma short-awned; rachilla extending beyond the base of the floret as a minute bristle. Fig. 172 . . . .
. . . . **WOODREED** *Cinna arundinacea* L.

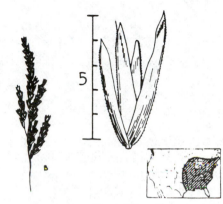

Figure 172

Perennial; culms in small tufts, 100-150 cm tall; panicles large, 15-30 cm long, drooping, the very flat spikelets shining, overlapping. The leaves are sometimes over 1 cm wide. This common woodland grass is found in moist forests. August-October.

Cinna latifolia (Trevir.) Griseb. has smaller spikelets, less than 4 mm long. Arctic America, southward through most of the United States except the southeastern states.

44. Alopecurus
Panicle dense and spikelike; spikelets freely dropping from the plant, the disarticulation below the glumes, which are equal, strongly keeled, and joined near the base; lemma with edges united near the base; palea absent; awn attached below the middle of the lemma.

1a Spikelets about 5 mm long; panicles 7-10 mm in diameter; awns protruding.

Fig. 173. . . . .
. . . . **MEADOW FOXTAIL**
Alopecurus pratensis **L.**

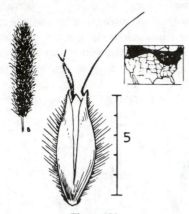

Figure 173

Perennial; tufted; culms 30-80 cm tall. The panicles resemble those of timothy but the lemmas have protruding bent awns and the spikelets fall off at a touch when ripe. This genus is one of very few having united glumes. Meadow foxtail is sometimes cultivated as a forage grass in the northern states and found growing wild in meadows and waste ground. Introduced from Europe. May-June.

1b Spikelets 2-3 mm long; panicles 4-5 mm in diameter; awns concealed in the glumes. Fig. 174. . . . .
. . . . *Alopecurus aequalis* **Sobol.**

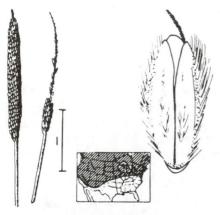

Figure 174

Perennial; culms weak and sprawling, 15-60 cm long; leaf blades 1-4 mm wide; panicles cylindrical, shattering very readily, 2-7 cm long. The plants are frequent on wet ground in swamps or along small streams. Succulent and probably good forage. May-June.

Alopecurus carolinianus Walt. has similar spikelets but the bent awn protrudes from the glumes. Tufted, erect annual. Throughout most of the United States. April-June.

45. Phleum
Panicle dense, spikelike, stiff; spikelets very flat and overlapping, fringed with short hairs; midnerve of each glume extended into a short, stiff awn; florets concealed within the glumes. Fig. 175 . . . .
. . . . **TIMOTHY** *Phleum pratense* **L.**

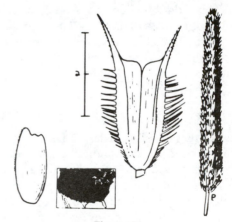

Figure 175

Perennial; tufted, the culms 50-100 cm tall, often with swollen, bulb-like bases. The dense, cylindrical panicles are stiff and somewhat bristly. Timothy is one of the important hay meadow grasses in the northern states, and is also very widely established in the wild in the moister portions of the country. June-July.

Phleum alpinum L. (ALPINE TIMOTHY), with short, plump, dark-colored panicles, occurs in wet mountain meadows at high altitudes in the West and from Greenland to Alaska.

46. Milium

Panicle open and drooping; glumes equal, covering and concealing the hard, shining, dorsally compressed floret; disarticulation above the glumes. Fig. 176 . . . .
. . . . *Milium effusum* L.

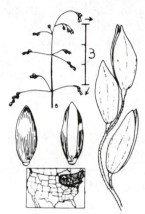

Figure 176

Perennial; culms in small tufts, up to 150 cm tall. Foliage smooth, the leaf blades 8-15 mm wide. The panicles are 10-20 cm long, very open, cylindrical, with short drooping branches. The dorsally compressed floret greatly resembles that of species of *Panicum,* but the disarticulation above the glumes places it with the Agrostideae. Cool, moist woods. May-July.

Tribe 8. Triticeae

47. Agropyron WHEATGRASS

Tufted or rhizomatous grasses; inflorescence a balanced spike; spikelets 1 at each node, laterally flattened against the rachis; florets several; disarticulation above the glumes and between the florets; lemmas acute or awned; grain usually retained in the floret.

1a Plants producing creeping rhizomes. . . . 2

1b Plants lacking rhizomes 3

2a Leaves bluish glaucous, often rolled; upper leaf surface furrowed, with 7-14 ridges across the width. Fig. 177 . . . .
. . . . WESTERN WHEATGRASS
***Agropyron smithii* Rydb.**

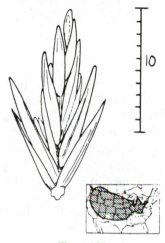

Figure 177

Perennial; culms 30-60 cm tall; leaves heavily glaucous, so that patches of the plants have a conspicuous blue or silvery color when viewed from a distance. Spikes slender, with erect spikelets. This is essentially a plant of the mountain and plains states of the West, but is occasionally found farther east. In the western parts of its range it is a source of forage on moist alkaline soil. Farther east, it grows on dry uplands, railroad embankments, etc. June-August.

2b Leaves flat, usually green; upper surface of blades finely nerved, with 25-40 nerves across the width. Fig. 178 . . . .
. . . . QUACKGRASS
***Agropyron repens* (L.) Beauv.**

Figure 178

Perennial; with extensive creeping rhizomes; culms 50-100 cm tall; leaves flat, green, often hairy on the upper surface; spike slender, with erect spikelets. Some of the plants have awned lemmas. Quackgrass is one of the worst grass weeds in the northern states. It produces good forage but is not intentionally planted as a crop. Waste ground, roadsides, fields and meadows. Introduced from Europe. May-July.

3a Spikelets erect, pressed against the rachis; rachis joints at least 5 mm long 4

3b Spikelets spreading away from the rachis, overlapping; rachis joints about 1 mm long. Fig. 179. . . . .
. . . . CRESTED WHEATGRASS
Agropyron desertorum (Fisch.) Schult.

Figure 179

Perennial; tufted; culms 60-100 cm tall. The strongly divergent spikelets distinguish this species from others of this genus. Crested wheatgrass is a recent introduction from the Old World and has proved very useful for regrassing abandoned crop lands and depleted ranges in the northern Great Plains. A valuable forage species. June-August.

Agropyron cristatum (L.) Gaertn. is called "Fairway crested wheatgrass." It is similar to the above, but the glumes are bent strongly to one side near their bases. Also cultivated.

4a Lemmas awnless or with straight awns. Fig. 180. . . . .
. . . . *Agropyron trachycaulum* (Link) Malte

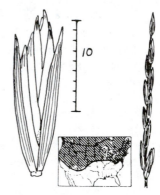

Figure 180

Perennial; tufted; culms 50-100 cm tall; spikes and spikelets slender; glumes conspicuously nerved. Awned and awnless forms occur, often in the same stand of plants. This is an important forage species in the western mountains. Moist grassland and open woods. As here discussed, this species includes forms sometimes called *A. pauciflorum* (awnless) and *A. subsecundum* (awned). June-September.

4b Lemmas with strongly bent awns. Fig. 181. . . . .
. . . . BLUEBUNCH WHEATGRASS
Agropyron spicatum (Pursh) S. & S.

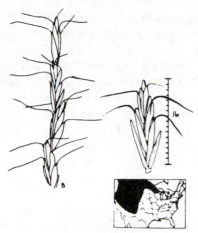

Figure 181

Wheat is the most extensively cultivated of the small grains and may be found growing in fields, waste places, and roadsides as the result of seed being scattered accidentally. Some varieties of wheat have awned lemmas, other are awnless. Two primitive kinds of wheat, EMMER and SPELT, have spikes with a brittle rachis which breaks into individual joints at maturity. Both are occasionally grown as feed grains in dry regions. All introduced from the Old World. Cultivated wheat was derived from ancient hybridizations of primitive wheats and and goatgrass *(Aegilops).*

Perennial; tufted; culms 60-100 cm tall. The spikes are slender and delicate. Sheaths smooth; blades 1-4 mm wide, hairy on the upper surface. This species is one of the most abundant and important forage species of the Northwestern States, and is readily eaten at all seasons by all kinds of livestock. Dry plains and open mountain slopes. May-August.

48. Triticum WHEAT
Tufted annual grasses; inflorescence a balanced spike; spikelets 1 per node; bracts blunt-tipped, short and broad, sometimes awned; disarticulation above the glumes and between the florets; grain in our species dropping from the floret when ripe. This genus is closely related to *Agropyron,* differing mostly in the floret shape and the free grain. Fig. 182. . . . .
. . . . **WHEAT** *Triticum aestivum* **L.**

49. Secale RYE
Cultivated tufted annual; spike slender; spikelets 1 per node; glumes short, awnless; florets 2, the rachilla prolonged beyond the second; lemmas strongly folded, with a row of short rigid bristles along the keel; disarticulation above the glumes; grain falling free at maturity. Fig. 183 . . . .
. . . . **RYE** *Secale cereale* **L.**

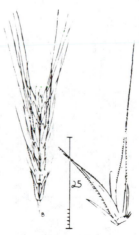

Figure 183

Figure 182

Rye resembles wheat in general appearance, but has a more slender spike. The lemmas are strongly V-shaped in cross section and have rows of vicious, upwardly pointed barbs along keel and margins. Rye culture is confined mostly to the northern states. The plants may be found scattered along roads and in waste places in regions where it is grown. Rye is often planted to stabilize the embankments of new roads. Introduced from Europe. Summer. Elongated blackish horns protruding from spikelets are ergot, a poisonous fungus used in medicine.

50. Aegilops

Annual weeds; inflorescence a cylindrical balanced spike; spikelets 1 per node, fitting against the rachis joint, which is thickened at its upper end; rachis disarticulating into individual internodes when ripe, each falling with the attached spikelet; florets 2-5; awns conspicuous. Fig. 184 . . . .
. . . . . **GOATGRASS** *Aegilops cylindrica* **Host**

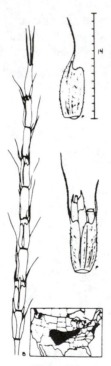

Figure 184

Annual; tufted; 40-60 cm tall; much branched from the base, one plant bearing as many as 60 spikes; spikes 5-10 cm long; joints of the rachis 6-8 mm long, the spikelets slightly longer and fitting closely into the contour of the joints;

spikelets glabrous or hairy, with 2-5 florets; lower spikelets nearly awnless, the upper ones bearing awns up to 5 cm long; glumes thick and stiff, bearing a pronounced tooth at one side of the awn. Goatgrass was presumably introduced into the Midwest in Turkey wheat brought to the United States by Russian immigrants in the 1870's. The plants are winter-annual, beginning growth in the fall and seeding out from May to July. They crowd out wheat when they are numerous. The seeds travel as contaminants in wheat seed.

Aegilops triuncialis L. is similar but has 3 awns on each glume. A bad weed on range land in California; introduced from Europe.

51. Elymus WILD RYE

Tufted or rhizomatous perennials or rarely annuals; inflorescence a balanced spike; spikelets usually 2 per node (rarely 1 or more than 2), placed laterally to the rachis, but the rachilla deformed so that the back of the first lemma is outward; florets 2 or more, usually awned; disarticulation above the glumes and between the florets.

1a **Spikes very broad; spikelets with 1 fertile floret and a rudiment; awn 5-10 cm long; nearly leafless annual. Fig. 185 . . . .**
. . . . **MEDUSA-HEAD** *Elymus caput-medusae* **L.**

Figure 185

This wiry annual weed has bristly spikes with very long awns. These may cause dangerous puncture wounds in grazing livestock. The plants are fibrous and furnish very poor forage, while they compete aggressively with better grasses. Northwestern states, to California. Introduced from Europe. Spring.

1b **Spikes much longer than broad; spikelets with several fertile florets; awns absent or less than 4 cm long; perennials, leafy . . . 2**

2a **Plants lacking rhizomes or rarely with short, thick ones** **3**

2b **Plants producing slender creeping rhizomes; ligule 1 mm long. Fig. 186.** . . . . ***Elymus triticoides* Buckl.**

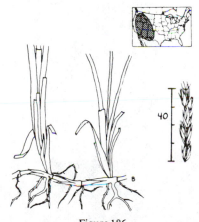

Figure 186

Perennial; culms usually 60-120 cm tall, single or in small tufts, forming large patches by means of the long, slender rhizomes. The leaves are harsh, stiff, bluish-glaucous and often rolled. Damp or saline soils, open ground. Sometimes part of the spike will have only a single spikelet at a node. Seven other species of *Elymus* produce extensive rhizomes. They are mostly plants of sand dunes along the oceans or in river valleys. May-August.

3a **Glumes very narrow, widest at the base and tapering upward, less than 1 mm broad** **4**

3b **Glumes flattened, wider near the middle than at the base, the middle portions with several nerves** **5**

4a **Lemmas awnless or with awns shorter than the body of the lemma; plants of the Great Plains and western mountains and deserts. Fig. 187** . . . . **GIANT WILD RYE** ***Elymus cinereus* Scribn. & Merr.**

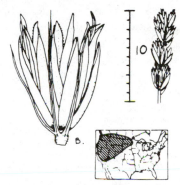

Figure 187

Perennial; tufted; rarely with short, thick rhizomes; culms about 1 m tall; leaves thick, stiff, flat or rolled; ligules 3-6 mm long; spikes 15-30 cm long. An important forage grass of dry plains, sand hills, and ditches in the western states. May-August.

Elymus condensatus Presl is a very tall species, up to 3 m high, with large, often compound spikes, up to 50 cm long and 3 cm thick. It grows along the Pacific Ocean beaches in California.

4b **Lemmas with awns at least as long as the body; plants of woodlands; eastern states and westward to Wyoming and Texas. Fig. 188.** . . . . ***Elymus villosus* Muhl.**

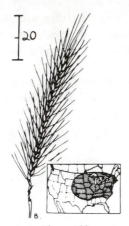

Figure 188

Perennial; tufted; culms slender, 60-100 cm tall; leaf blades thin and dark green, their upper surfaces velvety to the touch; spikes 5-12 cm long; spikelets usually densely hairy; glumes 12-20 mm long; lemmas 7-9 mm long, with an awn 1-3 cm long. Stream banks, thickets, and moist woods. June-August.

5a Spikes loose and curved; awns strongly recurved when dry; bases of glumes thin and flat. Fig. 189 . . . .
. . . . CANADA WILD RYE
Elymus canadensis L.

Figure 189

Perennial; tufted; culms erect or arching, bearing large, bristly spikes up to 25 cm long. The lemmas are rather coarsely hairy. Canada wild rye is our most widespread species of *Elymus* and is common over much of its range. It has been experimentally planted for forage production in the Midwest. Prairies, open ground, rocky banks and open woods. July-September.

Elymus riparius Wiegand is very similar but has awns which are straight even when mature and dry. The lemmas are scabrous, not hairy as in *E. canadensis*. Quebec to North Carolina, west to Nebraska and Arkansas.

5b Spikes stiff and straight; awns straight; bases of glumes round in cross section, hard and smooth. Fig 190 . . . .
. . . . VIRGINIA WILD RYE
Elymus virginicus L.

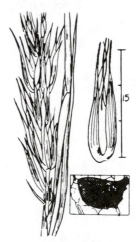

Figure 190

Perennial; tufted; culms stiffly erect, 60-120 cm tall; the base of the spike often hidden in the uppermost sheath. The "bowlegged" glumes are a good mark of recognition. Awned and awnless forms, as well as forms with smooth or hairy lemmas, are known. The plants are extremely variable in size and growth habit. The bases of the glumes are usually yellowish. Woods, thickets, stream banks, open ground. June-September.

52. Hystrix

Inflorescence a balanced spike of awned spikelets which spread at right angles to the rachis; spikelets 1-4 per node; glumes absent or reduced to short stubs; disarticulation below and between the florets. The genus is closely related to *Elymus*. Fig. 191. . . . .
. . . . **BOTTLEBRUSH**
Hystrix patula **Moench**

Figure 191

Perennial, in small tufts; culms slender, 60-120 cm tall. Spikes 8-15 cm long, very open, because of the spreading spikelets. Leaf sheaths smooth or hairy; blades usually 7-15 mm wide. This is a characteristic grass of damp woodlands throughout the eastern wooded section of the country. In var. *bigeloviana* (Fern.) Deam, the lemmas are hairy. June-August.

53. Hordeum BARLEY

Annual or perennial tufted grasses; inflorescence a balanced spike, the rachis (except in cultivated barley) breaking up at the base of each internode when ripe, and carrying at the apex of the internode a trio of spikelets; spikelets 3 at each node, the central one fertile, the lateral 2 usually reduced and sterile; floret 1, the rachilla of the central spikelet prolonged behind the palea.

1a Rachis of spikes separating at maturity into individual joints, each bearing 3 spikelets; wild plants. 2

1b Rachis of spikes not separating into joints; cultivated annual. Fig. 192. . . . .
. . . . **BARLEY** *Hordeum vulgare* **L.**

Figure 192

Annual; tufted; culms usually 60-120 cm tall; leaf blades flat, usually 5-15 mm wide, with prominent auricles; spikes stiff and erect; spikelets 3 at each node of the rachis; each spikelet with a single fertile floret; glumes awned; lemma with a stout awn, usually 10-15 cm long and very scabrous; rachilla prolonged behind the palea as a small bristle. In the hooded barleys, the lemmas lack awns but bear minute abortive extra lemmas at their tips, in inverted position. Common strains of barley are 6-rowed, that is, all 3 spikelets at each node develop grains. There are also 2 rowed barleys, in which only the central spikelet of each trio develops a grain. Barley is cultivated for the grain, which is used for human food (pearl barley, cereals), for production of malt, and for livestock feed. It is grown principally in the Midwest and in the Great Valley of California. Stray plants may grow from scattered seed, but barley is never found growing as a truly wild plant.

2a Awns mostly 5-8 cm long; plants perennial. Fig. 193. . . . .
. . . . FOXTAIL BARLEY
Hordeum jubatum **L.**

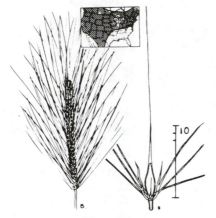

Figure 193

Tufted; culms 30-60 cm tall. The bushy spikes are 5-10 cm long, and because of the spreading awns, about as wide. When still fresh, the awns often have a purplish or pink sheen. When the spikes are ripe, they break up into segments, each with a sharp-edged rachis joint at its base, and a trio of spikelets at its apex. Only the central spikelet is fertile, and the lateral 2 are reduced to long awns. The spike-segments are able to penetrate clothing, wool, or flesh, causing puncture wounds, especially around the faces of animals. For this reason, foxtail barley is undesirable as a forage plant. Open ground, mostly in the midwestern and western states. May-August.

2b Awns less than 2 cm long; plants annual.

Fig. 194. . . . .
. . . . LITTLE BARLEY
Hordeum pusillum **Nutt.**

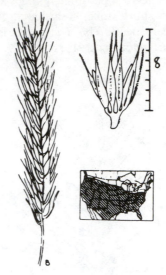

Figure 194

Annual; culms 10-35 cm tall, in small tufts. Both glumes of the central spikelet and the inner one of each lateral spikelet are broadened above the base. The plants furnish some early season forage, but are rejected by livestock after the spikes emerge. Plains and open ground; common in the Midwest and westward. February-July.

Hordeum brachyantherum Nevski is similar, but all of the glumes are narrow. Western mountain states, not common in the East.

Hordeum leporinum Link has thicker spikes which are nearly square in cross section and partially hidden in the uppermost sheath. Glumes of the central spikelet hairy on the edges. Mostly in the plains and mountain states. Introduced from Europe.

54. X Agrohordeum
Sterile hybrids of *Agropyron* and *Hordeum* species, having several florets per spikelet as in *Agropyron* and the disarticulating rachis of

Hordeum. Fig. 195 . . . .
. . . . **X *Agrohordeum macounii* (Vasey) Lepage**

Figure 195

This hybrid is commonly found on disturbed soil, along roads or in weedy pastures, where its parents, *Agropyron trachycaulum* and *Hordeum jubatum* grow together. The spikes commonly have 2 spikelets at the lower nodes. Above this there may be single spikelets with 3 glumes and normal single spikelets. No seed is produced, but the plants may gradually form sizable clumps. Iowa and Minnesota to Alaska and California. Sometimes called *Elymus macounii,* but not belonging to that genus.

55. X Elyhordeum

Sterile hybrids between various species of *Elymus* and *Hordeum,* having spikelets like those of *Elymus* but the disarticulating rachis of *Hordeum;* tufted perennials; inflorescence a balanced spike; spikelets 3 at each node, several-flowered, similar. Nine hybrids of this parentage have been found in various parts of North America. *Hordeum jubatum* is one parent of most of these, and they resemble this species in general aspect. Such hybrids should be looked for in disturbed sites where the parent species grow together.

56. Sitanion

Tufted perennials; inflorescence a balanced spike, the rachis disarticulating at the base of each internode; spikelets 2 at each node, several-flowered, long-awned, falling attached to the apex of the thin, sharp-edged rachis joint. Fig. 196 . . . .
. . . . **SQUIRRELTAIL**
Sitanion hystrix (Nutt.) J.G. Smith

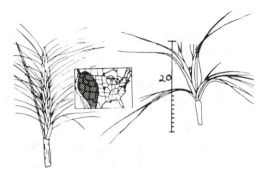

Figure 196

Culms 10-50 cm tall, erect or spreading. The glumes bear 1 or 2 long, bent awns. The plants resemble *Hordeum jubatum* but the several florets per spikelet and 2 spikelets per node separate it from that species. The rachis joints bearing the awned spikelets may penetrate the facial parts of grazing animals, causing serious inflammation, pink eye, etc. Fair forage when seed heads are not present. Dry woods and grasslands. April-September.

Sitanion jubatum J.G. Smith is similar but has 3 or more awns on each glume. Intermountain and Pacific coast states.

Tribe 9. Meliceae

57. Melica

Perennials, often with bulbous bases or short rhizomes; sheaths with united edges; inflorescence a panicle; spikelets disarticulating below the glumes or above the glumes and between the florets; upper florets sterile and enwrapping each other. Awned species are similar to *Bromus* but differ in having rudimentary upper florets.

1a **Spikelets disarticulating below the glumes and falling entire at maturity** **2**

1b **Spikelets disarticulating above the glumes and between the florets** **5**

2a **Rudimentary lemmas forming a pointed, cigar-shaped structure** **3**

2b **Rudimentary lemmas forming a blunt, club-shaped or bell-shaped structure** . . . **4**

3a **Glumes reaching nearly to the tip of the spikelet; spikelets V-shaped. Fig. 197**. . . . .
. . . . *Melica stricta* **Bolander**

Figure 197

Perennial; tufted; plants 15-60 cm tall; panicle slender, almost unbranched, raceme-like. All of the species of *Melica* have thin, translucent glumes and firmer lemmas. In this species, the glumes frequently have considerable purple coloration. Leaf blades scabrous, hairy on the top, 1-3 mm wide. Rocky and gravelly slopes in the mountains. Late May-August.

3b **Glumes less than 2/3 as long as the entire spikelet; spikelets narrow, cylindrical. Fig. 198**. . . . .
. . . . *Melica porteri* **Scribn.**

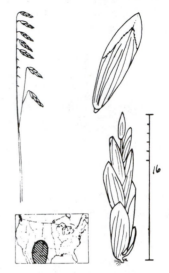

Figure 198

Perennial; tufted; 50-100 cm tall; inflorescences slender and raceme-like, 15-20 cm long, the branches ascending; spikelets mostly drooping to one side; pedicels hairy; spikelets green or brownish, not purple, 10-15 mm long, with 4-5 florets; lemmas strongly-nerved, minutely scabrous; sheaths smooth or scabrous, with united edges; leaf blades 2-5 mm wide. Open woods and slopes, moist ground; 2000-3000 m elevation. Midsummer-October.

, Var. *laxa* Boyle has spreading panicle branches 4-9 cm long; glumes sometimes purple. Western Texas (Chisos Mountains) to Arizona.

4a **Rudiment placed obliquely at the end of the rachilla; tips of fertile florets at the**

same height. Fig. 199. . . . .
. . . . *Melica mutica* Walt.

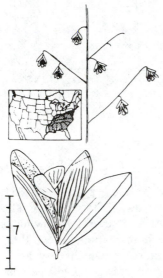

Figure 199

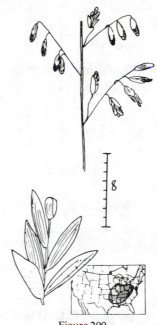

Figure 200

Perennial; tufted; plants 60-100 cm tall; panicles 10-20 cm long, with a few simple spreading branches bearing drooping spikelets. Sheaths hairy or scabrous; leaf blades 2-5 mm wide. Spikelets fan-shaped, 7-10 mm long, usually with 2 fertile florets and a bell-shaped rudiment, tilted sideways, at the end of the rachilla. This species, while seldom occurring in large stands, is the most widespread of the eastern *Melica* species. It grows in scattered stands in rocky woods. The name *Melica* refers to honey, but we do not know why Linnaeus applied it to this genus. April-June.

4b Rudiment placed straight on the end of the rachilla; tips of lower florets below that of the uppermost one. Fig. 200 . . . . .
. . . . *Melica nitens* (Scribn.) Hitchc.

Perennial; tufted; plants 50-120 cm tall; leaf sheaths glabrous or scabrous; leaf blades 7-15 mm wide; panicles 10-20 cm long, with a few branches. The drooping spikelets are more slender than in the previous species and usually have 3 fertile florets. The rudiment is more slender and pointed than in *M. mutica*. This species is probably more common than the previous one. The two are frequently confused but the position and shape of the rudiment should distinguish them. Scattered in rocky woods. Late April-June.

5a Lemmas without awns 6

5b Lemmas bearing awns. Fig. 201 . . . .
. . . . *Melica smithii* (Porter) Vasey

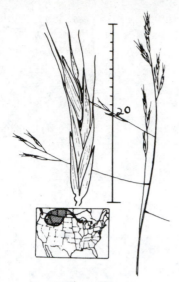

Figure 201

Slender tufted perennial; plants 60-120 cm tall; sheaths downwardly scabrous; leaf blades soft, scabrous, 6-12 mm wide. The panicles are very open, 12-25 cm long, with slender spreading branches bearing spikelets at the tips. The spikelets are 18-20 mm long, with 3-6 florets, and sometimes are purplish; awns 3-5 mm long. This slender woodland grass is found in moist forests. The occurrence in northern Michigan is the result of the cold post-glacial climate which once embraced that area. The plants resemble those of some of the perennial species of *Bromus,* from which they differ in the presence of the rudiment. July-August.

6a **Culms with bulbs at the bases** 7

6b **Culms without bulbs at the bases** 9

7a **Lemmas acute or obtuse, glabrous.** 8

7b **Lemmas tapering to an acuminate tip, pubescent. Fig. 202** . . . .
. . . . . . *Melica subulata* **(Griseb.) Scribn.**

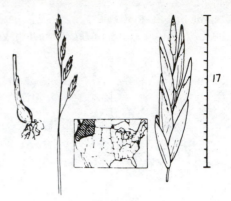

Figure 202

Perennial; tufted; plants 60-125 cm tall; panicles narrow, 10-15 cm long. The culms of this species and several following have small onion-like bulbs, about 1 cm long, at their bases, because of which they are sometimes called oniongrasses. Leaf blades usually 2-5 mm wide. Moist forests. May-July.

8a **First glume less than half as long as the spikelet; bulb attached to the crown of the plant by a thin stalk as much as 1 cm long. Fig. 203** . . . .
. . . . **ONIONGRASS**
Melica spectabilis **Scribn.**

Figure 203

Perennial; plants 30-100 cm tall; panicle slender, 10-15 cm long. This is a handsome species with somewhat inflated spikelets, the lemmas purple-tipped. Leaf sheaths hairy; blades flat or rolled, 2-4 mm wide. Gravelly mountain meadows and slopes. Forage value good. July-August.

8b First glume more than half as long as the spikelet; bulbs attached directly to a thick knotty crown. Fig. 204 . . . . ONIONGRASS
Melica bulbosa **Geyer**

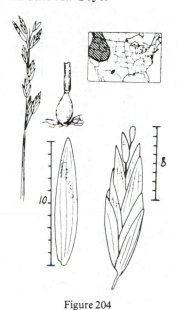

Figure 204

Perennial; plants 30-60 cm tall; panicle narrow and stiff. Leaf sheaths and blades smooth or hairy; blades 2-4 mm wide. This is perhaps the commonest western species in the genus. As the name implies, the plants usually have prominent bulbs at the bases of the culms. Woods and open slopes. Forage value good. July-August.

Melica fugax Boland. differs from the above species in having soft, thickish rachilla joints which turn tan and wrinkle when dried. The rachilla of *M. bulbosa* is thin, smooth, and

white. Panicles 8-15 cm long, with short, spreading or drooping branches. Dry ground, Washington to northern California and Nevada. Good forage for livestock and wild grazing animals. May-June.

9a Spikelets with 3 or more fertile florets . . . . go back to 8b

9b Spikelets with 1 or 2 fertile florets. Fig. 205 . . . . *Melica imperfecta* **Trin.**

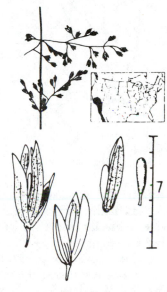

Figure 205

Perennial; tufted or with decumbent culms; plants 25-80 cm tall, bearing spreading panicles, 5-30 cm long, of numerous small, often purplish spikelets. In addition to the 1 or 2 fertile florets, there is a slender, yellowish rudiment which is 3-4 times as long as the very short rachilla joint which bears it. Gravelly soil. Good to excellent forage. April-May.

Melica torreyana Scribn. is similar but has hairy lemmas and a small rudimentary floret on a soft inflated rachilla half as long as the lemma. Central California.

58. Schizachne

Inflorescence a few-flowered panicle; spikelets disarticulating above the glumes; lemmas 7 nerved, awned between two teeth. Fig. 206 . . . . **Schizachne purpurascens (Torr.) Swallen**

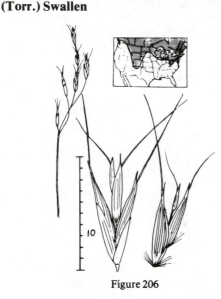

Figure 206

Perennial; tufted; 50-100 cm tall; sheaths with united edges; leaf blades narrowed at the base, 1-5 mm wide; panicles about 10 cm long, drooping. The glumes of the spikelets are usually purple. A rather delicate grass of rocky woodlands. It grows in scattered clumps and is seldom numerous. Also found in Japan and Siberia. May-July.

59. Glyceria MANNAGRASS

Tufted or rhizomatous perennials; leaf sheaths with united edges; spikelets very fragile, disarticulating early above the glumes and between florets; lemmas with 5-9 parallel nerves, usually blunt-tipped; plants of wet habitats.

1a Spikelets linear, usually 1 cm or more long, round in cross section, on short pedicels in narrow, erect panicles 2

1b Spikelets ovate or oblong, flattened, 7 mm or less long; panicles with drooping or erect branches 4

2a Lemmas obtuse; palea scarcely longer than the lemma 3

2b Lemmas acute, the palea much longer than the lemma. Fig. 207 . . . . *Glyceria acutiflora* Torr.

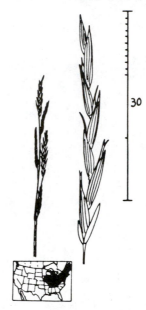

Figure 207

Perennial; rhizome-bearing; plants 50-100 cm tall; panicles slender, 15-36 cm long. Wet soil, swamps, or shallow water. The species of *Glyceria* all live in wet places, frequently in very shallow water. Their spikelets are very fragile and shatter at a touch when ripe. The plants are succulent and make good forage. May-August.

3a Lemmas 2.5-4 mm long, glabrous between the scabrous nerves. Fig. 208 . . . . *Glyceria borealis* (Nash) Batch.

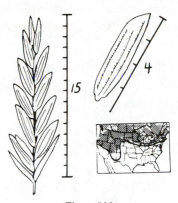

Figure 208

Perennial; culms erect or decumbent; plants 60-100 cm tall; panicles slender, erect, 20-40 cm long; leaf blades 2-6 mm wide; spikelets with 6-12 florets, 1-1.5 cm long. The inflorescence of this species is similar to that of the next (Fig. 209). Shallow water and marshy shores. June- September. This and the next species are very similar. The seeds of their close relatives in Europe are harvested from the water surface for human food.

3b Lemmas 4-5.5 mm long, minutely scabrous between the nerves. Fig. 209. . . . .
. . . . *Glyceria septentrionalis* Hitchc.

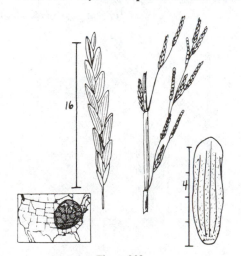

Figure 209

Perennial; culms spongy, 1-1.5 m tall, erect; panicles 20-40 cm long, with ascending branches; leaf blades 10-20 cm long, 4-8 mm wide; foliage smooth or the leaf blades minutely scabrous; spikelets 1-2 cm long, cylindrical, with 6-12 florets; lemmas about 4 mm long. A tall, succulent grass of shallow water and wet places, probably yielding good forage. May-July.

Glyceria fluitans (L.) R. Br. resembles the above but has lemmas 5-6 mm long, usually purple below the tip. Newfoundland to New York; South Dakota; Eurasia. In eastern Europe, the florets and grains of this species are harvested from the water surface for human food, being made into soup and gruel.

4a Panicle narrow, erect, the branches strongly ascending. 5

4b Panicle open, the branches drooping or spreading. 6

5a Lemmas about 3.5 mm long; panicle dense, oblong, about 15 cm or less long. Fig. 210. . . . .
. . . . *Glyceria obtusa* (Muhl.) Trin.

Figure 210

Perennial; culms erect or decumbent, 50-100 cm tall; leaf blades flat or folded; 2-6 mm wide; spikelets with 4-7 florets, 4-7 mm long. *Glyceria obtusa* is a characteristic grass of the bogs of the Atlantic Coastal Plain, found nowhere else. August-September.

5b Lemmas 2.0-2.7 mm long; panicle slender, 15-36 cm long. Fig. 211. . . . . *Glyceria melicaria* (Michx.) F.T. Hubb.

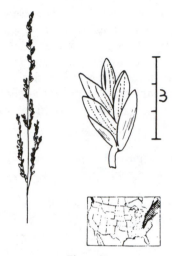

Figure 211

Perennial; culms in small tufts, 60-100 cm tall; leaf blades long and narrow, rough, 2-5 mm wide; spikelets with 3-4 florets, about 4 mm long. The slender, arching culms of this species fringe woodland streams and pools. July-August.

6a Spikelets oblong or ovate, mostly less than 2 mm wide; lemmas thin, with conspicuous nerves 7

6b Spikelets broadly ovate, 2-5 mm wide, with firm lemmas, the nerves not conspicuous. Fig. 212 . . . . *Glyceria canadensis* (Michx.) Trin.

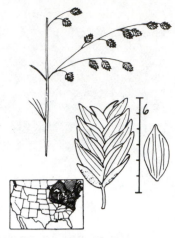

Figure 212

Perennial; tufted; plants 60-150 cm tall; panicles 12-20 cm long. *Glyceria canadensis* is one of the handsomest of grasses. The plump spikelets have a faint resemblance to snake rattles, hence the common name of "rattlesnake manna." Bogs, swamps, along streamlets. July-August.

7a First glume 1 mm or less long. 8

7b First glume 1.4 mm or more long. Fig. 213 . . . . *Glyceria grandis* S. Wats.

Figure 213

Perennial; tufted; culms thick and tall, 1-1.5 m high; leaf blades 6-12 mm wide; spikelets 4-7 mm long, with 4-7 florets; panicles 20-40 cm long. A tall species, with large, dense panicles, it is one of the commonest species of the genus within its range. Marshes and stream banks. June-August.

8a   Leaf blades 2—7 mm wide, firm; culms usually 1 m or less tall. Fig. 214 . . . .
. . . . *Glyceria striata* (**Lam.**) **Hitchc.**

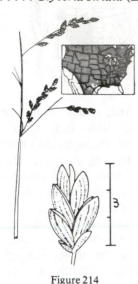

Figure 214

Perennial; forming large clumps; plants 30-100 cm tall; foliage glabrous; leaf blades flat or folded; panicles drooping, pyramidal, 10-20 cm long; spikelets 3-4 mm long, usually with 3-7 florets, often purplish. The spikelets shatter very readily when ripe. The most common and widespread of the American species of *Glyceria*. Usually one will find a colony of it around every pond, runnel, or ditch, and it may also grow in moist woods. Good forage for domestic livestock and elk, especially in later summer when the growth sites are drier. Sea level to 3300 m elevation. May-August.

8b   Leaf blades 6-12 mm wide, soft; culms usually 1-3 m tall. Fig. 215 . . . .
. . . . *Glyceria elata* (**Nash**) **Hitchc.**

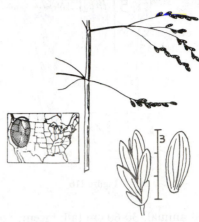

Figure 215

Perennial; tufted; dark green; tall and stout, with spongy culms; panicles oblong, 15-30 cm long, with spreading or drooping branches; foliage glabrous; leaf blades soft and thin; spikelets 4-6 mm long, with 6-8 florets. This species looks like a larger version of the previous one, but is restricted to the western states, where it is the most common and valuable forage species of the genus. Wet meadows, ponds, and moist woodlands. Eaten by all domestic livestock and elk; grazed primarily in late season when the herbage is less succulent and the habitats where it grows are less boggy. June-July.

60. Pleuropogon
Inflorescence a raceme of a few large spikelets; disarticulation above the glumes and between the florets; lemmas with 7 parallel nerves, usually awned; paleas winged or awned; sheaths with united edges. Fig. 216 . . . .
. . . . **SEMAPHORE GRASS** *Pleuropogon californicus* (**Nees**) **Benth. ex Vasey**

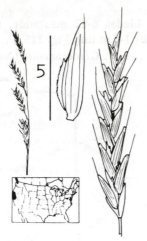

Figure 216

Tufted annual, 30-60 cm tall; raceme with 5-11 erect or spreading spikelets up to 2.5 cm long, each with 6-12 florets. Wet meadows and forests, northern coastal California. March-June.

Four similar species are found in the coastal region from central California to Washington.

Tribe 10. Brachyelytreae

61. Brachyelytrum
Panicle slender, arching; spikelets few, disarticulating above the glumes; first glume obsolete, the second minute; rachilla extending beyond the palea of the floret as a slender bristle. Fig. 217 . . . .
. . . . . *Brachyelytrum erectum* (Schreb.) Beauv.

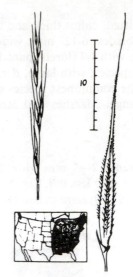

Figure 217

Perennial; tufted; culms 60-100 cm tall; leaf blades 10-15 mm wide, flat, rather light green; leaf sheaths and blades usually hairy; panicles slender, 5-15 cm long, with erect branches; first glume nearly absent; second glume up to 2 mm long; floret cylindrical, firm, about 1 cm long, with an awn 1-2 cm long. Typical *B. erectum* is found mostly south of the glaciated areas of the eastern United States. The lemmas have stout, stiff hairs along the nerves and are otherwise glabrous. Scattered stands in upland woods. June-July.

Var. *septentrionale* Babel grows mostly north of the glacial boundary, or southward in the mountains. The lemmas are sparsely and uniformly covered with minute appressed hairs or are nearly glabrous. Moist thickets and swampy woods. Also known from Japan.

Tribe 11. Stipeae

62. Stipa
Tufted perennials; inflorescence a panicle; glumes about as long as the floret; disarticulation above the glumes; floret rigid, cylindrical,

with a sharp hairy callus; lemma concealing the flat palea; awn strong, long, with a twisted basal segment. The sharp florets may injure livestock.

1a **Lemmas 5-12 mm long, excluding the awn** 2

1b **Lemma 15-25 mm long. Fig. 218** . . . . **PORCUPINE GRASS** *Stipa spartea* **Trin.**

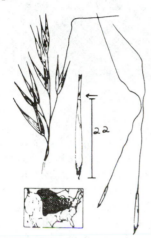

Figure 218

Perennial; culms in small hard tufts, about 1 m tall. The panicle is erect or nodding, with few spikelets; glumes whitish; mature lemmas brown; awns 15-20 cm long, with 1 or 2 sharp bends. The floret of *Stipa* species is a remarkable self-planting device. When it falls from the glumes, its sharp-pointed callus readily penetrates the ground. The backward-pointing hairs prevent the floret from pulling out. The twisted portion of the awn coils and uncoils as the moisture content of the air changes, causing the bent arm of the awn to revolve slowly until it comes in contact with grass stems or other objects. Then the whole lemma is literally screwed down into the earth. Unfortunately the same process will occur if the florets get into wool or hair of animals. These florets can thus cause serious puncture wounds on grazing animals, especially around the eyes, nose, and mouth. Prairies and dry open ground, mostly in the Midwest. June-July.

2a **Lemmas 8-12 mm long; glumes 12-20 mm long.** 3

2b **Lemma 5-6 mm long, chocolate brown. Fig. 219** . . . . *Stipa viridula* **Trin.**

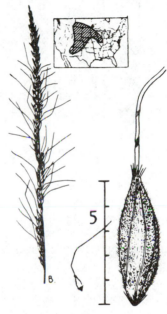

Figure 219

Perennial; culms 60-100 cm tall, in dense clumps; panicles slender, dense, elongated, 10-20 cm long. The glumes are glossy and rather translucent. Awn 2-3 cm long, with 2 bends. Dry plains. This species yields good forage. June-August.

Stipa robusta Scribn. (SLEEPY GRASS) is very similar, but taller (100-150 cm). It has a narcotic effect on horses who graze it. Dry plains, Colorado to Texas and Arizona.

3a Lemma with a smooth whitish cylindrical summit; awn with 2 sharp bends. Fig. 220 . . . . **TEXAS NEEDLEGRASS** *Stipa leucotricha* **Trin. & Rupr.**

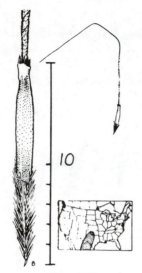

Figure 220

Perennial; culms 30-60 cm tall; leaf blades 10-30 cm long, very narrow, rolled, rough to the touch, dark green. The awn is 6-10 cm long, with 2 bends. The lower sheaths contain very peculiar hidden spikelets (cleistogenes) lacking glumes and with a very short-awned lemma. This species begins growth very early, in late winter and early spring and is prized for winter feed. The awned florets may injure sheep. Dry plains. May-June.

3b Lemma without a whitish ring; awn with only 1 bend, the upper segment curly. Fig. 221 . . . . **NEEDLE AND THREAD** *Stipa comata* **Trin. & Rupr.**

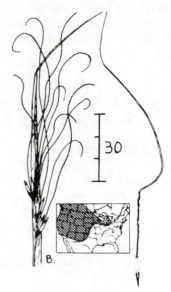

Figure 221

Perennial; culms in tufts, 30-60 cm tall. The panicle is a mass of curly awns, 10-15 cm long. The base of the panicle is usually partially hidden in the uppermost sheath. Needle and thread is a valuable forage grass in many parts of the West. It is grazed especially in the spring and fall, before the "needles" are formed, and after they fall. Prairies, plains, and dry open mountain slopes. June-August.

63. Oryzopsis

Tufted perennials; inflorescence a panicle; glumes equal, as long as the lemma; floret hard, oval-cylindrical, the lemma enclosing and concealing the palea; callus short, hairy; awn weak, not twisted, readily falling from the lemma. This genus is similar to *Stipa* but differs in the weak awns and plump florets. Occasional hybrids between species of *Stipa* and *Oryzopsis hymenoides* are known.

1a Lemmas covered with short, appressed hairs or glabrous 2

1b Lemmas covered with long, silky white hairs; panicle open, with spreading branches. Fig. 222 . . . . **INDIAN RICEGRASS** *Oryzopsis hymenoides* (R. & S.) Ricker

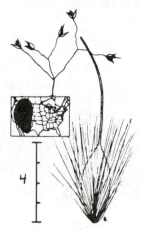

Figure 222

Perennial; culms 30-60 cm tall, in dense, tough tufts. The panicles are very open, with scattered spikelets on the tips of slender, zig-zag pedicels. The lemmas are brownish-black, but are covered with a dense cloud of white hairs. The awns readily break away from the lemmas, and may be missing from many of the spikelets. This is one of the most valuable forage grasses in the desert areas of the West. It is especially prized for winter feed. The large seeds are nutritious feed for livestock and formerly were used as food by the Zuñi Indians. April-August.

2a Spikelets (without the awns) 6-9 mm long; leaf blades flat, all at the base of the plant. Fig. 223 . . . . *Oryzopsis asperifolia* Michx.

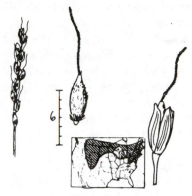

Figure 223

Perennial; tufted, with numerous long flat basal leaves; culms 20-70 cm long, usually spreading, lacking leaf blades (sheaths present). The hard cylindrical lemmas are yellowish. Scattered in sandy or rocky woods. May-June.

Oryzopsis racemosa (Smith) Ricker is similar in general habit, but has long upper culm leaves and short basal leaves. The lemmas are black. Quebec to Delaware, westward to Kentucky and South Dakota. June-August.

2b Spikelets (without the awn) less than 3 mm long; leaves threadlike; some leaf blades on the culms. Fig. 224 . . . . *Oryzopsis micrantha* (Trin. & Rupr.) Thurb.

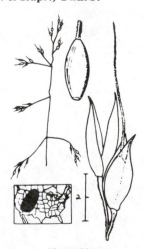

Figure 224

Perennial; in dense tufts; culms thin, 30-70 cm tall; leaf blades flat or rolled, less than 2 mm wide, scabrous; panicles 10-15 cm long, with slender spreading branches, the spikelets borne near the tips; glumes thin and translucent; lemmas smooth or with appressed hairs, yellow or brownish; awns readily detachable, straight, 5-10 mm long. Open woods and dry rocky slopes, intermediate elevations. The plants are said to have some forage value. June-July.

Oryzopsis miliacea (L.) Benth. (SMILO GRASS) has similar spikelets but broad, flat leaf blades, 8-10 mm wide. Cultivated for forage in California. Introduced from Europe.

64. Piptochaetium

Inflorescence a panicle; glumes equal, about as long as the rigid oval or cylindrical floret; lemma round in cross section, its edges not meeting but turned inward and fitting into a deep groove between the ridges of the palea, which is about as long as the lemma. Fig. 225 . . . .
. . . . . . **PIÑON RICEGRASS** *Piptochaetium fimbriatum* (H.B.K.) Hitchc.

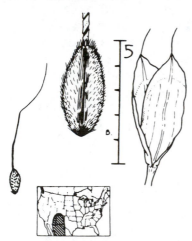

Figure 225

Perennial; culms in dense tufts, 40-80 cm tall; leaf blades 1/3 to 1/2 as long as the culms, mostly at the base of the plant, threadlike and curved downward; panicles 5-15 cm long; glumes about 5 mm long, thin; floret dark-colored, glabrous or hairy; awns readily

detachable, twice bent, 1-2 cm long. Piñon ricegrass is said to be a valuable forage species. Open, rocky woods.

The genus *Piptochaetium* is one of the oldest known grass genera. Fossils of the florets have been collected from Miocene rock formations in the western states.

The following two species have recently been shown to belong to this genus, although they have usually been assigned to the genus *Stipa*. Both have the turned-in lemma edges and ridged palea of *Piptochaetium*.

Piptochaetium avenaceum (L.) Parodi. Lemma brown, 9-10 mm long. Massachusetts to Michigan, southward to Florida and Texas.

Piptochaetium avenacioides (Nash) Valencia & Costas. Lemma brown, 15-20 mm long. Pine woods, Florida.

Tribe 12. Monermeae

65. Parapholis

Inflorescence a balanced cylindrical spike, the spikelets 1 at each node, fitting into a cavity which is closed off by the glumes; rachis disarticulating into single internodes when ripe, each segment containing a single-flowered spikelet. Fig. 226 . . . .
. . . . **SICKLE GRASS** *Parapholis incurva* (L.) Hubb.

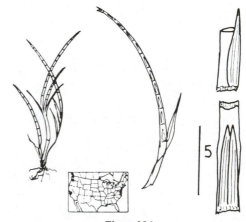

Figure 226

Low tufted annuals; clumps spreading; culms 10-20 cm long; spikes curved, 7-10 cm long, stiff. Dry banks and mud flats, salt marshes along the Atlantic and Pacific coasts; coastal Texas. Introduced from Europe. April-August.

Monerma cylindrica (Willd.) Coss. & Dur. is similar but the spikelets have only a single glume, placed directly in front of the single floret. Coastal California, from San Francisco Bay to San Diego.

Subfamily IV. Arundinoideae

Tribe 13. Arundineae

66. Arundo
Giant clump-forming grasses; panicles large, plumy; glumes slender, as long as the florets; disarticulation above the glumes and between the florets; lemmas 3 nerved, awned, covered with long hairs, giving the panicles a feathery appearance. Fig. 227 . . . .
. . . . **GIANT REED** *Arundo donax* L.

67. Phragmites
Tall rhizomatous marsh grasses, with plumy panicles; glumes shorter than the florets; lower florets longer than the upper ones; lemmas slender, awnless, 3 nerved, glabrous; disarticulation above the glumes and at the BASE of each rachilla internode; rachilla covered with long hairs, making the spikelets silky. Fig. 228 . . . .
. . . . . **REED** *Phragmites australis* (Cav.) Trin.

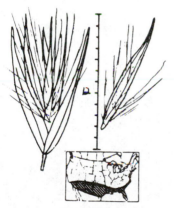

Figure 227

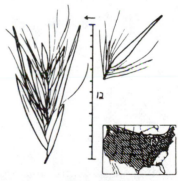

Figure 228

Perennial. This is one of the most spectacular of grasses of the temperate zone. The great culms reach a height of 6 m, and a thickness of 5 cm or more. The plume-like panicles produced by well-established clumps may reach a length of 60 cm. The stems do not persist over winter. Giant reed is a native of Europe, but is cultivated and naturalized in our southern states, and has proved hardy in cultivation as far north as central Iowa. The tough rind of the culms is used to make clarinet reeds. Fall.

Reed is a tall perennial grass, reaching 4 m or more in height, with smooth, polished stems and long and very broad leaves. The plants spread widely by vigorous rhizomes, forming great colonies along the margins of streams and in marshes and ditches. In autumn the large panicles become very feathery because of the hairy rachillas. The lowermost floret of the spikelet is staminate or sterile. Reed is widespread in the United States and is also known from all of the continents of the world. Fossil rhizomes of reed have been found in Europe, making it one of the few grasses known from past geological ages. July-October. Formerly called *Phragmites communis.*

68. Cortaderia
Giant clump-forming grasses with long, arching basal leaf blades; culms 2-7 m tall; inflorescence a large, plume-like panicle; plants dioecious; female spikelets with lemmas covered with long silky hairs; male spikelets glabrous. Fig. 229 . . . . **PAMPAS GRASS** *Cortaderia selloana* **(Schult.) A. & G.**

Figure 229

The giant plants are often grown as ornamentals in the southern half of the United States. Old plants make large, circular clumps with drooping, fountainlike basal leaves, their edges bristly, sawlike. The silvery panicles may be as much as 1 m long. They are often dyed and used in winter bouquets. Native of southern South America. Summer.

      Gynerium sagittatum (Aubl.) Beauv. This is a giant tropical grass, up to 10 m tall, native to the American tropics. It is probably not cultivated in the United States, but the enormous drooping panicles, up to 2 m long, often appear in store window displays. The plants are dioecious, the pistillate panicles being fuzzy, the staminate ones glabrous. Called Caña brava in Latin America.

Tribe 14. Danthonieae

69. Danthonia OATGRASS
Tufted perennials; spikelets few, in a panicle; glumes about as long as the spikelet, equal; florets usually 5 or more; lemma with a bent and twisted awn, arising between 2 teeth. The plants also produce hidden cleistogamous spikelets in the lower sheaths.

1a **Lemmas glabrous except for the hairy edges and callus; plants of the western mountains. Fig. 230.** . . . . *Danthonia californica* **Bolander**

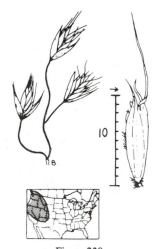

Figure 230

Perennial; tufted; culms 30-80 cm tall, tending to break at the joints. The panicle is small, consisting of 2-5 large spikelets on spreading pedicels. In this and other species of *Danthonia,* the lower sheaths may be swollen and contain slender, much distorted cleistogamous spikelets called "cleistogenes." The culms break off just below the node where a cleistogene is present. Meadows and open woods. Of some value for forage. May-August.

Danthonia intermedia Vasey has the few spikelets borne in a dense, tuft-like panicle, 2-5 cm long. Arctic North America, southward to Michigan and at high altitudes in western states. Grazed by livestock. July-September.

1b **Lemmas hairy on the back as well as the edges; plants widespread in the United States. Fig. 231** . . . .
. . . . **POVERTY OATGRASS**
Danthonia spicata (L.) Beauv.

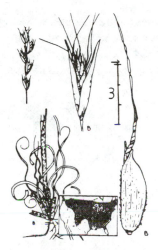

Figure 231

Perennial; tufted; culms usually 20-50 cm tall. Most of the short, curly leaves are borne in a basal tuft. Their ligules are under 1 mm long. Panicles short, 2-5 cm long, with ascending branches. Poverty oats makes gray-green mats or sods on dry, sterile soils in the open or in thin woods. Forage value very low. Cleistogenes may be present, as in *D. californica*. May-September.

Danthonia compressa Austin has panicles with spreading branches. The ligules of the lower leaves are 2-5 mm long. Open ground and in moist or dry woods, mostly in the Appalachian Mountains; Quebec to Georgia. June-July.

70. Schismus
Small tufted annual grasses; spikelets disarticulating above the glumes and between the florets, or occasionally the whole spikelets falling with a stub of the pedicel attached; glumes and lemmas many-nerved; glumes nearly or quite as long as the entire spikelet; lemmas blunt or acute, with 2 short teeth at the tip, sometimes with a short awn between them; grain obovoid, golden, translucent, dropping from the floret. Fig. 232. . . . .
. . . . *Schismus barbatus* (L.) Thell.

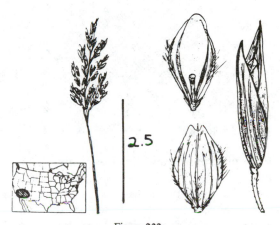

Figure 232

Schismus grass is a winter annual weed in the southwestern deserts at low altitudes. Most of the leaves are basal. The plants are dwarf, 5-20 cm tall, but furnish some winter forage. Lower florets about 1.8-2.2 mm long. Introduced from the Mediterranean. March-May.

S. arabicus Nees is similar but has lemmas over 2.5 mm long with acute teeth at the tip, sometimes with a minute awn between them. Arizona to California. Spring.

Tribe 15. Chasmanthieae

71. Chasmanthium
Perennial woodland grasses; spikelets much flattened, the 2 many lemmas strongly keeled;

glumes short; lower 1-4 lemmas empty; lemmas awnless, 5-15 nerved; disarticulation above the glumes and between the florets; anther 1. Fig. 233 . . . . *Chasmanthium latifolium* (Michx.) Yates

Figure 233

Perennial; producing rhizomes; plants 1-1.4 m tall; panicles open and drooping, 10-20 cm long. While similar to sea oats, this is a woodland species of rich soil in the southeastern United States. It is highly ornamental and suitable for growing in shaded borders or wild gardens. Several other species of this genus with smaller spikelets also occur in the southeastern states. June-October.

This species has been usually assigned to the genus *Uniola*. While the spikelets are similar, it differs in anatomical characters and disarticulation.

Tribe 16. Aristideae

72. Aristida NEEDLEGRASS, WIREGRASS
Tufted annuals and perennials; inflorescence a panicle; spikelets 1 flowered, disarticulating above the glumes; floret hard, cylindrical, the lemma enwrapping the flower; callus sharp, hairy; awns 3, the lateral 2 often shorter than the central one, which may be coiled. There are about 40 species found in the United States. They are especially common on dry soil in the southern and western states. Most are indicators of overgrazing or soil disturbance. They yield little forage and are dangerous to livestock.

1a   Central awn spirally coiled at the base . . 2

1b   Central awn not coiled. 3

2a   Glumes about equal in length; lateral awns about 1/4 as long as the central one. Fig. 234. . . . . *Aristida dichotoma* Michx.

Figure 234

Annual; tufted, in small, shallowly-rooted clumps; culms 20-40 cm tall. The panicles are very slender, almost racemose. The species of *Aristida* are sometimes called wiregrasses or needlegrasses. They are typically grasses of depleted ranges or poor, sterile soils. This species is found on dry sandy or rocky open ground or in open sterile woods. Forage value negligible. August-October.

2b Glumes unequal, the first 1/2 to 3/4 as long as the second. Fig. 235 . . . .
. . . . *Aristida basiramea* **Engelm.**

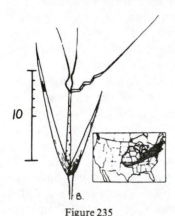

Figure 235

Annual; culms in small tufts, 30-50 cm tall; leaf blades harsh, 1-2 mm wide; panicles slender, raceme-like, 5-10 cm long, at the tips of the culms and in axils of the upper leaf sheaths. Dry sterile soil, open ground. August-October.

Var. *curtissii* (Gray) Shinners has very short lateral awns, 2-4 mm long. Pennsylvania and Virginia to Wisconsin, Wyoming, Colorado and Arkansas.

3a Glumes about equal, 2-3 cm long; lemma about 2 cm long. Fig. 236 . . . .
. . . . *Aristida oligantha* **Michx.**

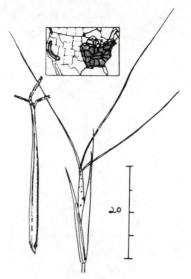

Figure 236

Annual; tufted; culms much branched, 30-50 cm tall; leaf blades narrow, usually under 1 mm wide; panicles 10-20 cm long; glumes nearly equal, 2-3 cm long, sometimes 3-cleft at the tip; awns 4-7 cm long, spreading, about equal in length. The plants are wiry and almost leafless, most of the height being the bristly inflorescences. The sharp-pointed, 3-awned florets of this and other species of *Aristida* are great "crawlers." The pointed and backwardly-barbed callus penetrates hair or clothing easily, and every movement of the body of the host results in the floret digging in deeper, aided by the scabrous awns. Forage value negligible. Dry open ground. August-October.

3b Glumes unequal, the first about 1 cm long, the second twice as long; lemma 12-15 mm long. Fig. 237 . . . .
. . . . **DOGTOWN GRASS**
Aristida longiseta **Steud.**

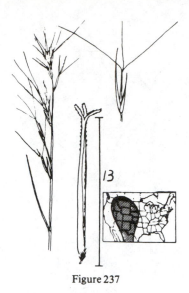

Figure 237

Perennial; in large tufts; culms 20-50 cm tall; panicles narrow, standing well above the leaves, appearing as a feathery mass of long, reddish awns. The awns may reach lengths of 6-8 cm. The narrow, stiff, straight leaves may be mostly at the base of the plant. Dogtown grass is an inferior forage species. The sharp awns and pointed callus cause the florets to pierce the facial tissues of grazing animals, causing serious infections and sometimes blindness. Dry plains and foothills, especially on thin rocky soil or bare ground. July-October.

Aristida fendleriana Steud. has similar panicles and spikelets, but most of the leaves are crowded in dense, curly basal tufts. Widespread in the Great Plains and western mountain states.

Subfamily V. Chloridoideae

Tribe 17. Diarrheneae

73. Diarrhena

Rhizomatous perennial; culms in small tufts; inflorescence a simple panicle of few spikelets; spikelets several-flowered, disarticulating above the glumes and between the florets; lemmas stiff, 3 nerved; grain when ripe flask-shaped, protruding from the floret, thick-walled and rigid, the seed free from the wall. Fig. 238 . . . .
. . . . *Diarrhena americana* **Beauv.**

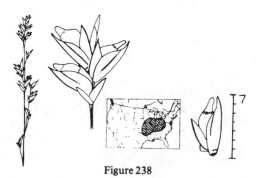

Figure 238

Perennial; producing numerous scaly rhizomes; culms up to 1 m in height; leaves mostly low on the culm, the blades 1-2 cm wide, scabrous or hairy; panicle slender, with short erect branches, nodding, 10-30 cm long. Spikelets 10-18 mm long. Growing in scattered clumps in rich woods. The peculiarly shaped grains are unique among our grasses. July-October.

Recent studies by C. Schwab indicate that this genus belongs to the Chloridoideae.

Tribe 18. Unioleae

74. Uniola

Coarse rhizomatous or stoloniferous grasses of marine sand dunes; inflorescence a panicle of large, strongly compressed and keeled spikelets that disarticulate below the 3 nerved glumes; lower 4-6 florets empty; lemmas awnless, with many faint nerves; anthers 3. Fig. 239 . . . .
. . . . **SEA OATS** *Uniola paniculata* **L.**

Figure 239

Figure 240

Perennial; producing strong rhizomes; plants up to 1 m tall; panicles dense, nodding, 20-40 cm long. The very flat spikelets are strikingly ornamental. Sea oats inhabits the coastal sand dunes along the Atlantic and Gulf of Mexico and the West Indies. The panicles are frequently harvested and used for winter bouquets and display window ornaments. Spikelets may be found on the plants at most seasons of the year.

Tribe 19. Aeluropodeae

75. Distichlis
Tough rhizomatous dioecious perennials, growing in salty or alkaline habitats; panicles small, with few, crowded spikelets; spikelets of staminate and pistillate plants similar; disarticulation above the glumes and between the 5-15 florets; nerves of the strongly keeled lemmas numerous but faint. Fig. 240 . . . .
. . . . **SALT GRASS**
Distichlis spicata (L.) Greene

Perennial; spreading by stiff, scaly rhizomes; plants 10-40 cm tall. Coarse, stiff plants of saline or alkali flats in the drier parts of the western states and in coastal salt marshes. In desert areas, they are readily eaten by cattle, but are seldom taken if better forage is available. April-September.

Var. *stricta* (Torr.) Beetle is the interior form of the species, occurring through much of the western states.

Var. *spicata* occurs in coastal marshes along the entire Atlantic, Gulf, and Pacific coasts of the United States. It is somewhat larger and more vigorous than the inland form.

Tribe 20. Eragrosteae

76. Eragrostis LOVEGRASS
Annuals or perennials, usually tufted; inflorescence a panicle, sometimes with spikelike branches; spikelets small, with 3 many florets; glumes short; lemmas 3 nerved; disarticulation above the glumes and between the florets, or in many species the glumes and lemmas falling from a persistent rachilla.

1a Plants forming flat mats, the trailing culms rooting at the nodes 2

1b Plants erect or with somewhat decumbent culms, but never rooting at the nodes. . . 3

2a Staminate and pistillate spikelets on separate plants; anthers about 1.5-2 mm long; panicle dense. Fig. 241 . . . . *Eragrostis reptans* (Michx.) Nees

Figure 241

Annual; stoloniferous; only a few cm tall, forming delicate, bright green turf along streams and on wet ground. Spikelets borne in dense panicles, resembling clover heads. Lemmas very strongly keeled. While the two sexes are similar in appearance, they can be distinguished by the presence of anthers or stigmas protruding from the florets. Forage value low. Late summer and fall. Also called *Neeragrostis reptans*.

2b Spikelets all with perfect flowers; anthers minute, about 0.2 mm long. Fig. 242 . . . . *Eragrostis hypnoides* (Lam.) B.S.P.

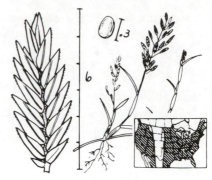

Figure 242

Annual; stoloniferous, forming low mats. The panicle is usually open. Lemmas very strongly keeled. The anthers are minute and nearly round, with a bulk of less than 1/50 of the anthers of *E. reptans*. Like most of its relatives, this plant is a warm-weather grass, growing rapidly from midsummer on. Stream banks and wet ground. Forage value low. July-September.

3a Lemmas when mature dropping from the persistent rachilla of the spikelet; (this can be detected by pulling lemmas outward and downward) paleas usually remaining on the rachilla. Fig. 243 15

Figure 243

3b Lemmas not falling separately; spikelets disarticulating between the florets when mature. **4**

4a Paleas without long fringing hairs **5**

4b Paleas fringed with long straight hairs which nearly cover the spikelets. Fig. 244. . . . . *Eragrostis ciliaris* (L.) R. Br.

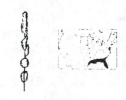

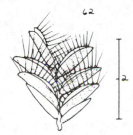

Figure 244

This delicate tufted annual grass grows 15-30 cm tall, and bears dense cylindrical panicles, like those of the foxtail grasses, 3-10 cm long. The tiny spikelets, only 2-4 mm long, are immediately identifiable by the long, fringe-like cilia borne on the margins of the palea, which give a spider-web appearance to the spikelets. River banks and open ground. Widespread in warm regions of the world. June-August.

5a Spikelets 3-15 mm long **6**

5b Spikelets 2-3 mm long; panicle elongated, dense, narrowly cylindrical. Fig. 245 . . . .

. . . . *Eragrostis glomerata* (Walt.) Dewey

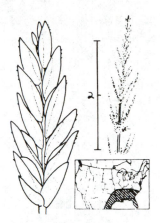

Figure 245

Annual; tufted; plants 1 m or less tall. *E. glomerata* produces a profusion of slender, cylindrical panicles up to 50 cm long, with strongly ascending branches. The plants flower from near ground level to the very tip. The tiny spikelets have very thin, translucent lemmas. At maturity the grains are visible through the lemmas. Banks of streams and ditches, alluvial woods. July-November. Sometimes called *Diandrochloa glomerata*.

6a Spikelets sessile or nearly so, strongly keeled and flattened. **7**

6b Spikelets on slender pedicels; lemmas keeled or rounded on the back **8**

7a Spikelets few, distant along the few elongated panicle branches, a sessile spikelet and a cottony tuft of hairs in the axil of each panicle branch. Fig. 246 . . . . *Eragrostis sessilispica* Buckl.

Figure 246

Perennial; tufted; plants usually 20-40 cm tall. This is a tumbleweed grass, the panicle breaking off and rolling with the wind when mature. The leaves are clustered in a short basal tuft. The panicle may reach as much as 40 cm in length. The main axis is somewhat spiral; the branches are straight, stiff, and bear sessile spikelets. Dry sandy plains. Forage value apparently low. May-June.

7b Spikelets numerous, in a dense panicle; no sessile spikelets nor cottony tufts in the axils. Fig. 247 . . . .
. . . . *Eragrostis secundiflora* **Presl**

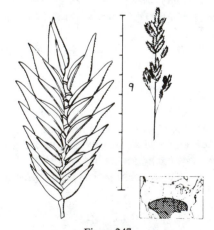

Figure 247

Perennial; wiry; tufted; plants 20-70 cm tall; leaves 1-5 mm wide, smooth except at the base of the blade. The spikelets are borne in 1 or more dense tufts along the axis of branches of the panicle. The spikelets often overlap like shingles. They usually have a reddish or bronzy color when ripe. Sandy or rocky open ground. Forage value apparently low. May-December. Other names that have been used for this species are *E. oxylepis* and *E. beyrichii*.

8a Panicle branches stiffly spreading; spikelets deep reddish-purple; lemmas strongly keeled. Fig. 248 . . . .
. . . . **PURPLE LOVEGRASS**
Eragrostis spectabilis **(Pursh) Steud.**

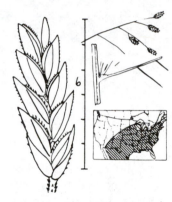

Figure 248

Perennial; tufted; erect or rarely spreading, up to 60 cm tall. Purple lovegrass has a large, open, dome-shaped panicle which makes up about 2/3 of the height of the plant. The axils of the panicle branches and throats of the sheaths bear conspicuous tufts of white hair. The leaf sheaths may be smooth or hairy. This is one of our most widespread grasses on sandy open ground, and one of the most attractive. July-September.

8b Panicle branches delicate, not stiffly spreading; lemmas not strongly keeled. . 9

9a Sheaths glabrous, or pubescent on only the upper edges 10

9b Sheaths pubescent along the edges, on the surface, and on the collar. 11

10a Lemmas 1.8-2.4 mm long; panicle broad, ovoid. 14

10b Lemmas 2.4-3.4 mm long; spikelets 4 mm or more long; panicle elongated, ellipsoid. Fig. 249. . . . .
. . . . SAND LOVEGRASS
Eragrostis trichodes (Nutt.) Wood

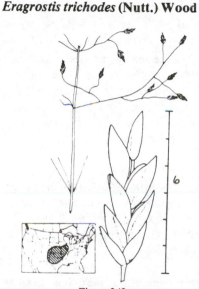

Figure 249

11a Sheaths and often blades bearing hairs, each of which arises from a little blister. Fig. 250 12

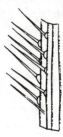

Figure 250

11b Hairs of sheaths not arising from little blisters. 13

12a Spikelets 5-10 mm long; lemmas 2.4-3.4 mm long . . . . . go back to 10b

12b Spikelets 2-5 mm long; lemmas 2.0-2.4 mm long. 13

13a Panicle not over twice as long as wide; usually 15 cm wide or wider 14

13b Panicle 3 times as long as wide; never over 12 cm wide; cultivated and sometimes escaping. Fig. 251. . . . .
. . . . WEEPING LOVEGRASS
Eragrostis curvula (Schrad.) Nees

Perennial; tufted; 80-120 cm tall. The open, cylindrical panicles may make up half the total height of the plant. The spikelets frequently have purplish florets and yellow glumes. Sandy plains and open woodlands. Sand lovegrass has high forage value, but has been virtually destroyed by overgrazing in many of the areas where it once abounded. August-September.

Eragrostis curvula (Fig. 251) sometimes has smooth sheaths and might be keyed out here. It may be recognized by the lead-colored, short-pedicellate spikelets.

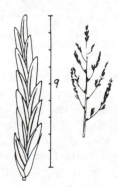

Figure 251

Perennial; tufted; 60-120 cm tall, forming large round clumps. The leaves are long, drawn out to very fine drooping tips. The panicles are elongated and somewhat drooping; branches not bearing spikelets near the bases; spikelets dull straw colored or leaden gray. Native to South Africa, weeping lovegrass was first brought to the United States as an ornamental, but is now widely planted in the southern states as a forage grass for revegetating abandoned or eroded crop land. It makes excellent pasturage and can be used for hay. The seeds are about 1 mm long, smooth, amber colored except for the blackish germ. Summer.

Eragrostis trichodes (see Fig. 249) sometimes has hairy sheaths and may key out here. It may be recognized by the long-pedicellate, usually bronzy or yellowish spikelets.

14a   Lemmas 2.0-2.4 mm long; larger leaf blades 5-10 mm wide. Fig. 252 . . . .
. . . . *Eragrostis hirsuta* (Michx.) Nees

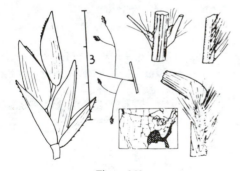

Figure 252

Perennial; tufted; plants becoming 1 m or more tall; panicles large, open, cylindrical, half the height of the plant. The leaves are wider than in most other species of this genus and taper to long fine points. The summits of the leaf sheaths (see figure) vary from extremely hairy to glabrous. Dry soil, in open woods and fields. Summer.

14b   Lemmas 1.8-2.2 mm long; larger leaf blades 2-3 mm wide. Fig. 253 . . . .
. . . . *Eragrostis intermedia* Hitchc.

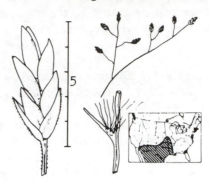

Figure 253

Perennial; tufted; reaching nearly 1 m in height. This species has an open, dome-shaped panicle with slender delicate branchlets. The leaf blades are narrow, involute, and drawn out to slender tips; sheaths glabrous or the lower ones somewhat hairy. Dry, sandy, open ground. June-September.

15a   Plants perennial, usually 50-150 cm tall, forming large clumps from hard, knotty bases; basal buds of next year's growth present at flowering time . . . .
go back to 6

15b   Plants annual, usually less than 50 cm tall, from soft, shallow-rooted bases; no basal buds present at flowering time. . . 16

16a   Spikelets with 2-4 (rarely 5) florets 17

16b   Spikelets with 6 many florets (rarely 5 in starved plants) 18

17a   Pedicels of *lateral* spikelets 3—many times as long as the spikelets; panicle large, diffuse, 2/3 or more of the total

height of the plant; plants erect. Fig. 254 .
. . . . **LACEGRASS**
Eragrostis capillaris (**L.**) **Nees**

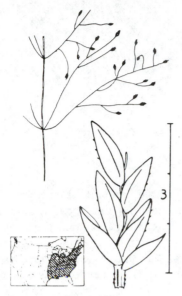

Figure 254

Figure 255

Lacegrass is a densely tufted annual, up to 50 cm tall. Culms much branched at the base, the plants bearing numerous panicles; leaf sheaths somewhat hairy, especially at the throat; blades flat, hairy on the upper surface, 1-3 mm wide; panicles open-cylindrical or elliptical, making up most of the height of the plants. Spikelets 2-3 mm long, with 2-4 florets; glumes about 1 mm long; lemmas about 1.5 mm long. The minute grains are about 0.5 mm long. A common weed of open, dry situations on waste ground, in fields, and thin woods. August- September.

17b   Pedicels of *lateral* spikelets short, rarely more than twice the length of the spikelets; panicles about half the length of the sprawling plants. Fig. 255 . . . .
. . . . *Eragrostis frankii* **C.A. Mey.**

Annual; a weed of river banks and wet alluvial bottoms, *E. frankii* makes sprawling bushy tufts. The culms are seldom more than 25 cm long. Leaf sheaths and blades usually smooth except at the throat. The ellipsoidal panicle is much denser than that of *E. capillaris,* because of the shorter pedicels. August-September.

18a   Plants without glands on spikelets or branches; spikelets 2 mm or less wide . . 19

18b   Plants bearing minute blister-like glands on the keels of the glumes and lemmas and the branches of the panicle; spikelets 2.5-3.5 mm wide when mature. Fig. 256. . .
. . . . **STINKGRASS**
Eragrostis cilianensis (**All.**) **Lutati**

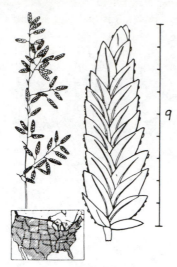

Figure 256

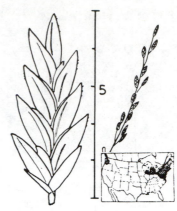

Figure 257

Annual; tufted; culms erect or somewhat spreading; panicles ovoid or pyramidal, rather dense. The keels of the glumes and lemmas bear tiny circular glands. Stinkgrass is a vigorous weedy annual, with a strong, musty odor when fresh. It may be poisonous to horses if eaten in large quantities. Frequent in fields, gardens, and dry, disturbed soil. Introduced from Europe and now very common throughout the United States. Also known as *E. megastachya.* June-October.

A very similar species, *E. poaeoides* Beauv., has spikelets with glands on the keels of the lemmas also. The spikelets range from 1.3-2.0 mm wide. It is less common than stinkgrass. Introduced from Europe.

19a Sheaths bearing a few long hairs on the margins at their summits; paleas remaining on the rachilla **20**

19b Upper sheaths (and usually the lower ones as well) lacking long hairs at their summits; panicle rather dense, the lower branches usually single, bearing 11-40 spikelets; paleas falling from the rachilla with the lemmas. Fig. 257 . . . . *Eragrostis multicaulis* **Steud.**

Weedy, tufted, annual, usually under 30 cm tall. Spikelets 3-4 mm long, usually with 4-8 florets. After the paleas fall, the minute zigzag rachillas remain as the only evidence of the spikelets. This species grows mostly as a weed in cities in the northeastern states. Although introduced in the Americas, its homeland is not known with certainty. Also known as *E. peregrina.* July-October.

20a Spikelets mostly lying closely appressed to the panicle branches. Fig. 258 . . . . *Eragrostis pectinacea* **(Michx.) Nees**

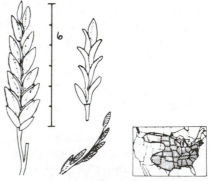

Figure 258

Weedy annual; tufted; culms usually 20-30 cm tall; plants branching freely from the base, forming dense, erect tufts. Spikelets 5-8 mm long. The manner in which the spikelets lie parallel to the panicle branches is characteristic. Frequent on dry roadsides, waste ground, and cultivated fields; one of the commonest weedy annual grasses. July-October.

20b Spikelets when mature diverging strongly from the panicle branches; spikelets 0.7-1.4 mm wide; lateral nerves of lemmas obscure. Fig. 259 ....
.... *Eragrostis pilosa* (L.) **Beauv.**

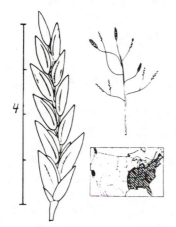

Figure 259

Annual; tufted; delicate; culms 10-50 cm tall; panicles open, ellipsoidal. The tiny spikelets, 3-5 mm long, stand away from the main panicle branches on hair-like pedicels. Leaf blades 1-3 mm wide, flat. Introduced from Europe. July-September.

77. Leptochloa

Perennials and annuals; inflorescence a panicle of 1-sided racemes, the spikelets on very short pedicels, appressed in 2 rows to the lower sides of the branches; spikelets with several awned or awnless florets; lemmas 3 nerved; disarticulation above the glumes and between the florets.

1a Spikelets 5-10 mm long; lemmas awned or awnless **2**

1b Spikelets 1-2 mm long; lemmas awnless. Fig. 260 ....
.... **RED SPRANGLETOP**
Leptochloa filiformis (Lam.) **Beauv.**

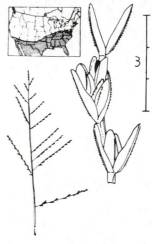

Figure 260

Annual; plants tufted, 20-130 cm tall. The reddish or purple panicles may be half the height of the plants, with numerous threadlike spikes bearing the very tiny spikelets. Red sprangletop is a rather rampant weed, frequent on bottomlands, in fields, and in gardens. August-September.

2a Lemmas notched at the blunt apex, awnless, glabrous or nearly so. Fig. 261 ..
.... **GREEN SPRANGLETOP**
Leptochloa dubia (H. B. K.) **Nees**

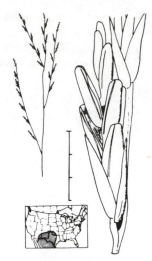

Figure 261

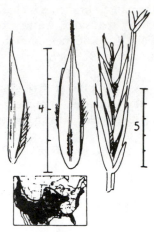

Figure 262

Perennial; tufted; culms 50-100 cm long, tough, erect; leaf sheaths smooth; blades flat or somewhat rolled or folded, up to 10 mm wide; panicles up to 15 cm long, the spreading spikes 3-12 cm long; spikelets 5-10 mm long, usually with 5-8 or occasionally fewer florets; lemmas oblong, blunt, the tip notched, the midnerve sometimes protruding. The plants sometimes bear cleistogamous inflorescences hidden in the sheaths. This species has some value as a forage grass in the Southwest. Sandy or rocky open ground. March-September.

2b Lemmas tapering to a sharp point, awned, hairy on the nerves. Fig. 262 . . . . *Leptochloa fascicularis* (Lam.) Gray

Annual; tufted; culms 30-100 cm long, erect or horizontally spreading; plants becoming much-branched; leaf blades flat or somewhat rolled; panicles rather stiff, usually partly hidden in the leaf sheaths, 10-20 cm long, the individual spikes up to 10 cm long; spikelets 7-12 mm long, with 6-12 florets; awns ranging from very short to 4-5 mm long. Moist or alkaline soil, salt marshes, open ground. June-September. This is an unusually wide-ranging grass, extending southward through the American Tropics to Argentina. It is sometimes placed in the genus *Diplachne*.

78. Eleusine
Inflorescence of several radiating spikes; spikelets in 2 rows along the lower side of the rachis, disarticulating above the glumes and between the florets; florets several, flattened and keeled, the 3 nerves prominent, all close together. Fig. 263 . . . .
. . . . **GOOSEGRASS**
Eleusine indica (L.) Gaertn.

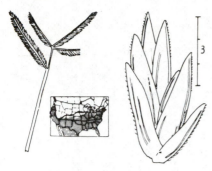

Figure 263

Annual; culms up to 50 cm long, spreading horizontally or standing erect; plants often making flat mats. Goosegrass is a very common weed of fields, gardens, paths, and disturbed ground generally in the southeastern United States. Introduced from the warmer sections of the Old World. March-October.

79. Dactyloctenium

Inflorescence of several radiating spikes, the rachis extending beyond the spikelets; spikelets in 2 rows along the lower side of the rachis, flattened and keeled; second glume with a short, curved awn. Fig. 264 . . . .
. . . . *Dactyloctenium aegyptium* (L.) Willd.

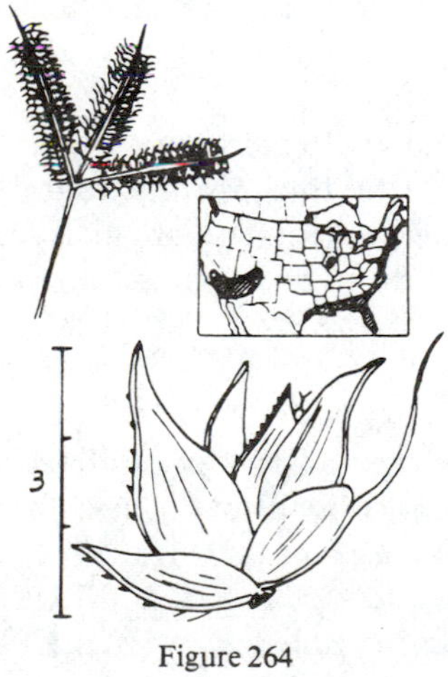

Figure 264

Annual; tufted; culms upright or spreading out and forming mats, rooting at the lower nodes.

The seeds (ovary wall is lost) are brownish, about the size of a pinhead, and oddly sculptured. Weed on cultivated ground. This species was apparently once planted by the Indians along the lower Colorado River for grain. Introduced from Europe. Summer, or almost yearlong in the far South.

80. Tridens

Perennials; inflorescence an open or cylindrical panicle; leaf blades flat, not white-margined; spikelets with short or long glumes and several florets; lemmas split or toothed at the tip, with 3 prominent nerves, their tips often protruding as short points; disarticulation above the glumes and between the florets; stigmas purple.

1a   **Lemmas whitish, glabrous. Fig. 265** . . . .
. . . . *Tridens albescens*
(Vasey) Woot. & Stand.

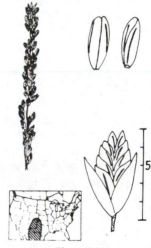

Figure 265

Perennial; tufted or rarely with rhizomes; culms 30-80 cm tall; panicles spikelike; leaf blades elongated, 2-4 mm wide, sometimes inrolled. This species may be confused with species of *Eragrostis,* but the fact that the lemma is split at the tip excludes it from that genus. The plants often grow in ravines and around water holes. They are succulent and make good forage, but are seldom found in dense stands. Also known as *Triodia albescens*. April-October.

1b Lemmas brownish or purple, hairy 2

2a Panicle open, with spreading, drooping branches. Fig. 266 . . . . PURPLETOP
Tridens flavus (L.) Hitchc.

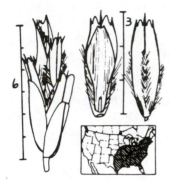

Figure 266

Perennial; tufted; culms 1-1.5 m tall; panicles graceful and open, up to 35 cm long; leaf blades smooth, flat, 3-10 mm wide. The 3 nerves of the lemma protrude as little points. Open or partially shaded grassy places. Purpletop is eaten by livestock to some extent. It is attractive and colorful and might be grown as an ornamental. Also known as *Triodia flava*. August-October.

Var. *chapmanii* (Small) Shinners is similar. It has a stiffer and more open panicle. A circle of hairs extends completely around the base of each main and secondary panicle branch. Dry woods, mostly on the Coastal Plain, New Jersey to Texas, north to Iowa.

2b Panicle slender and spikelike. Fig. 267 . . . . *Tridens muticus* (Torr.) Nash

Figure 267

Perennial; tufted; culms 30-50 cm tall. The spikelets often have a faint purplish hue before drying. Leaf blades very narrow, about 1 mm wide. Overgrazed lands and dry rocky slopes. Forage value low. Also known as *Triodia mutica*. June-October.

Tridens elongatus (Buckl.) Nash is very similar, but has leaf blades up to 3-4 mm wide; plants 40-80 cm tall; panicles 10-25 cm long. Missouri to Texas and Arizona.

81. Erioneuron

Low perennials with culms of 1 or 2 internodes; leaf blades with thick, white edges; panicles small and dense; lemmas long-hairy along the margins and the midrib; paleas long-hairy near the base.

1a Plants spreading by stolons; panicles surrounded by a tuft of leaf blades; lemmas lobed at the tip. Fig. 268 . . . . FLUFFGRASS
Erioneuron pulchellum (H.B.K) Tateoka

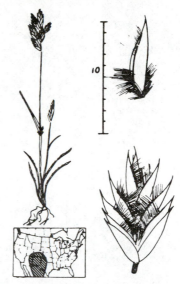

Figure 268

Figure 269

Perennial; tufted at first, then spreading by short stolons; usually less than 15 cm tall. The entire plant consists of a tuft of thread-like basal leaves, from which arise culms which have a single long internode and bear a cluster of spikelets and leaves at the summit. These culms soon bend over and root at the tip. This process may be repeated until a mat of the plant is built up. Fluffgrass is found on thin soils, overgrazed lands, and deserts. Forage value negligible; usually regarded as an indicator of overgrazing. Also known as *Tridens pulchellus* or *Triodia pulchella*. April-October.

1b Plants tufted; panicles long-stalked, not surrounded by leaves; lemmas not lobed. Fig. 269 . . . .
. . . . . *Erioneuron pilosum* (Buckl.) Nash

Perennial; tufted; 10-30 cm tall. Most of the very narrow white-margined leaves are at the base of the plants. The culm usually consists of a single internode. The plants are very shallow-rooted and easily pulled up. Found frequently on thin rocky soils and overgrazed ranges of the Southwest. This species has little forage value and is generally regarded as an indicator of overgrazing. Also known as *Tridens pilosus* or *Triodia pilosa*. March-October.

82. Triplasis
Panicles small, partly concealed in the uppermost sheath; spikelets several-flowered; lemmas blunt, 2 lobed, the 3 parallel nerves hairy, the central one protruding as a short awn; palea fringed with long hairs on its upper half. Fig. 270 . . . .
. . . . *Triplasis purpurea* (Walt.) Chapm.

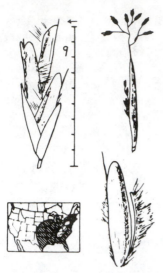

Figure 270

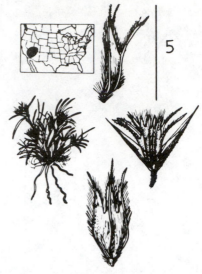

Figure 271

Annual; tufted; culms 30-75 cm long, erect or spreading. In late season, the herbage may become quite reddish. The small terminal inflorescence usually protrudes somewhat from the uppermost sheath, but the axillary panicles are concealed in the swollen sheaths and have cleistogamous spikelets. Common on sandy lands. July-October.

Triplasis americana Beauv. is a similar species of the southeastern states. It is perennial and the lemmas have awns about as long as the body.

83. Blepharidachne
Dwarf tufted perennial; inflorescence a dense panicle; glumes as long as the spikelet; disarticulation above the glumes, the florets falling as a group of 4, only the third one fertile; lemma with 3 nerves, all hairy and projecting as short awns; lemma split into 2 lobes along the midrib. Fig. 271 . . . .
. . . . *Blepharidachne kingii* (Wats.) Hack.

These diminutive desert plants are usually less than 10 cm tall. They resemble *Erioneuron pulchellum,* but do not make stolons. Only the third floret contains a flower. Deserts from Utah to California. Spring.

84. Redfieldia
Rhizomatous sand-binding perennials; inflorescence an open panicle; glumes short; florets several; lemmas 3 nerved, with a tuft of long hairs on the callus; disarticulation above the glumes and between the florets. Fig. 272. . .
. . . . **BLOWOUT GRASS**
Redfieldia flexuosa (Thurb.) Vasey

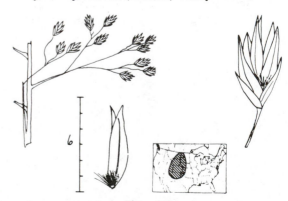

Figure 272

Perennial; culms 60-100 cm tall; panicles large, pyramidal; leaf blades smooth and tough, elongated, inrolled. The spikelets when mature are fan-shaped, with conspicuous cottony tufts visible from the side. Because of the numerous long, slender rhizomes, blowout grass is able to keep pace with the shifting sands and eventually bind dunes into place. An important sand-binding grass on sandy plains from South Dakota to Oklahoma and Arizona. August-October.

Certain forms of *Diarrhena americana* (Fig. 238) may key out here. *Diarrhena* is a grass of rich, moist woods, with soft, flat leaf blades.

85. Scleropogon
Dwarf stoloniferous perennial; dioecious; staminate inflorescence of awnless, non-disarticulating spikelets; pistillate spikelets disarticulating only above the glumes, the florets falling together. Fig. 273 . . . .
. . . . **BURRO GRASS**
Scleropogon brevifolius **Phil.**

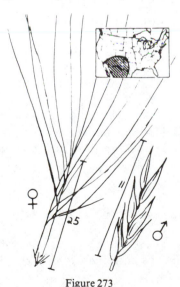

Figure 273

Burro grass is a low perennial, 10-20 cm tall, spreading by stolons and making patches. Staminate and pistillate inflorescences on the same or different plants. The inflorescences are small tufts of a few erect spikelets. Each pistillate spikelet has a number of florets, which are shed from the glumes as a unit, covered with numerous awns. Each lemma looks much like a floret of *Aristida,* with three long awns. The sharp awns and callus of the pistillate lemmas can penetrate hair, wool, and facial tissues of grazing animals. Burro grass tends to replace more desirable grasses on overgrazed lands of the arid Southwest. Forage value very low. June-September.

Tribe 21. Sporoboleae

86. Muhlenbergia
Tufted or rhizomatous grasses; inflorescence an open or spikelike panicle; spikelets small, 1 flowered, disarticulating above the glumes; glumes shorter than the floret or with awn-tips extending beyond it; floret soft; lemmas 3 nerved, awned or awnless, the callus usually hairy.

1a   Glumes minute, less than 1/5 as long as the lemma. Fig. 274 . . . .
. . . . **NIMBLE WILL**
Muhlenbergia schreberi **Gmel.**

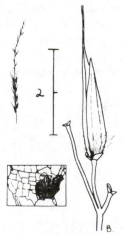

Figure 274

Perennial. The culms in early season are quite erect, but by flowering time they become much-branched and sprawl on the ground, the lower nodes often rooting. The erect portions of the culms are 10-30 cm long. Leaf blades usually 2-4 mm wide; foliage glabrous; panicles borne at the tips of the culms and from leaf axils, slender and weak, 5-15 cm long. The glumes are vanishingly small, the first sometimes entirely lacking and the second only a few tenths of a millimeter long. Florets cylindrical, about 2 mm long, hairy on the callus; awn 2-5 mm long, very slender. Nimble Will may become a weed in shaded lawns and shrubby borders, but it does not seem very aggressive. Also found growing in woods and thickets, roadsides and city streets, old fields, and meadows. August-October, rarely blooming in June or July.

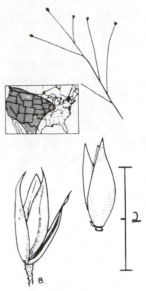

Figure 275

1b **Glumes at least half as long as the lemma**.........................2

2a **Plants producing elongated usually scaly rhizomes**.......................3

2b **Plants lacking rhizomes (old tufts sometimes stooling out)**............7

3a **Panicles slender or spikelike, with short ascending branches; spikelets awned or awnless, on short pedicels**...........4

3b **Panicles open, with spreading branches; tiny awnless spikelets on long slender pedicels. Fig. 275**...................
...............**SCRATCHGRASS**
Muhlenbergia asperifolia (N. & M.) Parodi

Perennial; bushy, 10-50 cm tall, with rhizomes; panicles open, dome-like, 5-15 cm long; plants pale green; leaf blades 2-5 cm long, 1-2 mm wide; ligules minute. The grains are often swollen into spherical shape by the action of a fungus. Moist, often alkaline soil. June-September.

Muhlenbergia arenacea Buckl. is similar but has prominent ligules, 1-2 mm long, Texas to Colorado and Mexico.

4a **Leaf blades 3 mm or more wide, more than 5 cm long, flat and thin; lemmas awned or awnless**..................5

4b **Leaf blades 1-2 mm wide, less than 5 cm long, usually rolled; lemmas awnless. Fig. 276**...........................
...................**MAT MUHLY**
Muhlenbergia richardsonis (Trin.) Rydb.

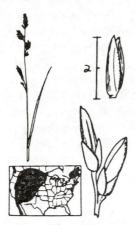

Figure 276

Perennial; much-branched, spreading by hard, thin rhizomes and forming mats, the erect portions of the culms 5-60 cm long; panicle slender, 2-10 cm long. Mat muhly is a grass of open, often wet or alkaline soil. While rather tough, it furnishes fairly good livestock feed. The densely matted plants furnish considerable erosion protection to the soil. Also known as *M. squarrosa* (Trin.) Rydb. July-September.

5a Glumes much longer than the awnless floret, tapering to awn-tips; panicle dense, spikelike, bristly. Fig. 277. . . . . . . *Muhlenbergia racemosa* **(Michx.) B.S.P.**

Figure 277

Perennial; culms in bushy tufts, with numerous erect branches, plants 30-100 cm tall; rhizomes densely covered with overlapping scales; internodes of the culms smooth and shiny. Anthers 0.5-1.0 mm long. Prairies, dry or moist open ground; Mississippi Valley and westward, from southern Canada to the southwestern states. July-October.

M. glomerata (Willd.) Trin. is similar, but has large anthers 1-1.5 mm long and dull, minutely hairy internodes. Marshes and wet soil; northeastern and north central states; transcontinental in southern Canada. July-October.

5b Glumes mostly shorter than the awned or awnless floret, lacking awn-tips; panicle slender. 6

6a Culms much-branched, smooth and shining; axillary panicles produced at many nodes and mostly partly hidden in the sheaths. Fig. 278 . . . . *Muhlenbergia frondosa* **(Poir.) Fern.**

Figure 278

Perennial; rhizomes present; culms becoming elongated, 40-100 cm or more long, the plants becoming bushy and much-branched, frequently scrambling through bushes or other vegetation or sprawling; panicles at the tips of the culms and protruding from nearly every leaf sheath, up to 10 cm long, rather dense; leaf blades flat, scabrous, 3-7 mm wide; glumes 2-4 mm long, tapering gradually from base to an awned tip; lemmas 2-3 mm long, awnless or awned. This is a very common species in thickets and woods and on roadsides and stream banks. August-October.

Muhlenbergia bushii Pohl (*M. brachyphylla* Bush) is very similar, but has glumes considerably shorter than the lemma; ligules very short, about 0.5-0.7 mm long. Mississippi Valley states.

6b   Culms sparingly branched; panicles at the tip of the culm and of elongated erect axillary branches; culms dull, the internodes usually covered with minute hairs, especially below the nodes. Fig. 279 . . . .
. . . . . . *Muhlenbergia mexicana* (L.) Trin.

Figure 279

Perennial; in tufts, producing abundant scaly rhizomes; plants often becoming bushy; panicles lobed; spikelets in dense clusters, the pedicels very short; glumes 2-3 mm long, tapering gradually from the base to a short awn

point, about as long as the body of the lemma; floret hairy near the base, about 3 mm long. Awned and awnless plants occur in the same colony. Marshes, moist shores, open moist woodlands. July-October.

Muhlenbergia sylvatica Torr. is similar but has a more slender and open inflorescence, some of the spikelets being on elongated pedicels. The ligules are over 1 mm long. Northeastern United States and southern Canada, mostly east of the Mississippi River. August-October.

7a   Second glume not toothed 8

7b   Second glume 3 toothed near the tip. Fig. 280 . . . .
. . *Muhlenbergia montana* (Nutt.) Hitchc.

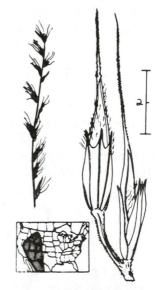

Figure 280

Perennial; culms in large, dense tufts, 30-60 cm tall; panicles slender, with ascending branches. The 3 toothed second glume is the best identifying mark of this species. The old sheaths at the base of the plants become flat and stiff, like thin wooden splints. This species yields fairly palatable forage, especially when the herbage is young. Ponderosa pine, spruce, and fir forests, 2300-3300 m elevation. July-October.

8a **Panicle very narrow, the short branches bearing spikelets nearly to their bases. . . 9**

8b **Panicle broad, open, the spikelets borne near the tips of the branches. 10**

9a **Awn shorter than the lemma or lacking. Fig. 281. . . . .
. . *Muhlenbergia cuspidata* (Torr.) Rydb.**

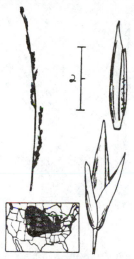

Figure 281

Perennial; tufted; culms slender and wiry, 20-40 cm tall; ligules minute; panicles very slender. Dry hills and prairies. July-September.

Muhlenbergia wrightii Vasey (SPIKE MUHLY) is similar but has a denser panicle, somewhat like timothy. The ligule is 1-2 mm long. It is an important grazing grass on open or bushy ranges, ponderosa pine forests, from southern Colorado and Utah southward.

9b **Awn 1-several times as long as the lemma. Fig. 282. . . . .**

. . . . .BULLGRASS
Muhlenbergia emersleyi Vasey

Figure 282

Perennial; in large tufts; culms tall and stout, 50-100 cm tall; sheaths glabrous, flattened and keeled; lower leaf blades up to 50 cm long; blades flat or folded, rough, 1-4 mm wide; ligules thin, 1-2 cm long; panicles long and narrow, 20-40 cm long, with ascending, overlapping branches; spikelets often somewhat purplish; glumes scabrous; lemmas hairy on the lower half, awnless or with an awn up to 25 mm long attached below the apex. Some panicles may have both awned and awnless spikelets. Canyons and rocky woods. Said to be a good soil binder. September-October.

10a **Plant forming sprawling, much-branched bushes from knotty crowns. Fig. 283. . . .**
. . . . BUSH MUHLY
Muhlenbergia porteri Scribn.

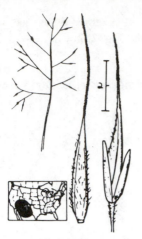

Figure 283

Perennial; the much-branched plants will form bushy growth 1 m in diameter and height if ungrazed. Because of the very high palatability of this species for grazing livestock, plants are rarely seen except in the hearts of spiny bushes. The plants are semi-evergreen, the old culm bases producing new shoots in the succeeding year. Dry plains and deserts. July-September.

10b Stems erect, usually unbranched. Fig. 284 . . . .
. . . . **RINGGRASS**
Muhlenbergia torreyi **(Kunth) Hitchc.**

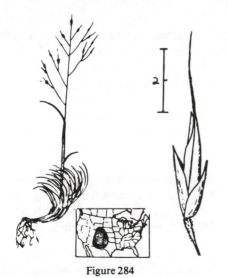

Figure 284

Perennial; plants forming circular or ring-shaped tufts, with numerous short, curly leaves, mostly clustered at the base of the plant. Ringgrass is a rather poor forage grass. Its presence usually indicates that better species have been killed out. Open plains, 1300-3300 m elevation. July-August.

87. Lycurus
Inflorescence a dense, spikelike panicle; spikelets single-flowered, in pairs which drop from the rachis as units; first glume with 2 or 3 awns; floret awned, hairy. Fig. 285 . . . .
. . . . **WOLFTAIL**
Lycurus phleoides **H.B.K.**

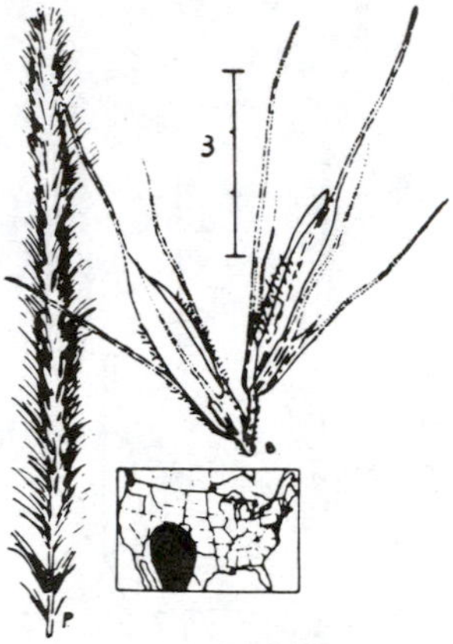

Figure 285

Perennial; tufted; culms 20-60 cm tall. The dense, spikelike, bristly panicles are usually gray. The spikelets fall in pairs, one of each pair being on a longer stalk than the other. The lower spikelet is staminate and the upper one perfect-flowered. Open brushy hillsides and ponderosa pine forests, 1400-2600 m elevation. A valuable forage grass, grazed especially in spring. July-September.

88. Sporobolus DROPSEED
Inflorescence an open or dense panicle; spikelets small, 1 flowered, the glumes usually shorter than the single floret; lemma awnless, 1 nerved; palea conspicuous; caryopsis when ripe readily slipping from the floret, its ovary

wall becoming thick and gelatinous when wet (except in a few species).

1a   Glumes plainly unequal in length......2

1b   Glumes equal in length, nearly as long as the floret; plants annual, in small tufts with very shallow roots. Fig. 286.......
. . *Sporobolus vaginiflorus* (Torr.) Wood

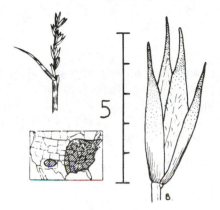

Figure 286

Annual; culms thin, wiry, 10-40 cm tall, in small tufts; panicles mostly concealed within the upper leaf sheaths, or only the tips protruding. The lemmas are usually blackish-spotted, and the palea is often longer than the lemma. Both lemma and palea are covered sparsely with appressed hairs. Dry, sterile, open ground. August-October.

Sporobolus neglectus Nash is similar, but has shorter, plumper spikelets with glabrous lemmas. Northeastern and midwestern United States.

2a   Spikelets 4-7 mm long..............3

2b   Spikelets 1-2.5 mm long.............4

3a   Panicle open, with spreading branches. Fig. 287.........................
............PRAIRIE DROPSEED
Sporobolus heterolepis Gray

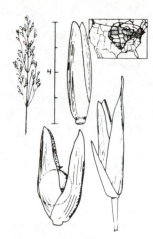

Figure 287

Perennial; plants forming large tufts; basal leaves elongated, arching and drooping; culms 30-70 cm tall; panicles narrowly ovoid. The spikelets become much distended by the ripening grain, which is spherical and yellowish at maturity and about 2 mm in diameter. The palea readily splits down the middle as the grain develops. Prairies. July-October.

3b   Panicle spikelike, mostly hidden in the uppermost sheath. Fig. 288...........
......*Sporobolus asper* (Michx.) Kunth

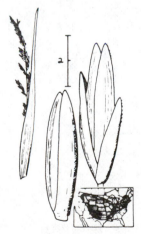

Figure 288

Perennial; tufted; culms 60-120 cm tall; leaf blades flat or rolled, 1-4 mm wide, tapering to a slender tip; panicles whitish or somewhat purplish in color, 5-15 cm long. Open ground and prairies. August-September.

Var. *macer* (Trin.) Shinn. is very similar but has scaly rhizomes. Pine forests; Mississippi to Oklahoma and eastern Texas.

Var. *drummondii* (Trin.) Vasey is a form with a slender panicle. Iowa to Texas and Mississippi.

Sporobolus clandestinus (Biehl.) Hitchc. is similar but has hairy florets; lemma and palea slender-pointed, the palea longer than the lemma. Dry sandy lands; Connecticut to Wisconsin, Kansas, Texas and Florida.

4a **Sheaths glabrous or nearly so at the summit** **5**

4b **Sheaths bearing conspicuous tufts of white hairs at their summits. Fig. 289** . . . . **SAND DROPSEED** *Sporobolus cryptandrus* **(Torr.) Gray**

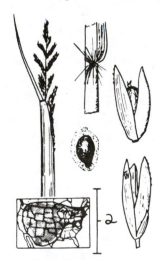

Figure 289

Perennial; tufted, the erect or spreading culms 30-100 cm long; panicles borne at the apex of the culms and in the axils of the upper sheaths. The tufts of straight silky hairs on the flanges at

the summits of the sheaths are prominent. Sand dropseed is a rather important forage species on coarse or sandy soil in the West. It produces an abundance of fine, long-lived seed, and will recover rapidly from the effects of overgrazing. July-October.

Sporobolus giganteus Nash (GIANT DROPSEED) resembles the above, but may be distinguished by its large size, 1-2 m tall, and slightly larger spikelets, 2.5-3 mm long. Sand plains, western Oklahoma and Texas to Colorado and Arizona.

5a **Panicle not more than twice as long as wide; leaf blades usually rolled. Fig. 290** . . . . **ALKALI SACATON** *Sporobolus airoides* **Torr.**

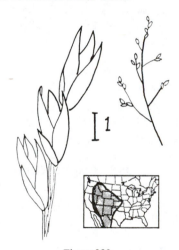

Figure 290

Perennial; culms tough and rigid, in large clumps; plants 50-100 cm tall. The large, open, dome-shaped panicles make up half or more of the height of the plant. The paleas of the florets often split as the grain develops. The plants can grow on very salty or alkaline soil. Although the herbage is tough, it is taken greedily by livestock, probably because of the large amount of salt in the tissues. Under a hand lens, minute salt crystals frequently can be seen glistening on the leaf surfaces. Plains and alkali flats, often on heavy clay soils. June- September.

5b Panicle 3 or more times longer than wide; leaf blades usually flat. Fig. 291.......
....................SACATON
Sporobolus wrightii **Munro**

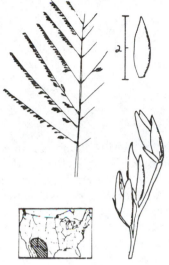

Figure 291

Perennial; culms in large tufts, stout and tough, 1-2.5 m tall; panicles up to 60 cm long, the branches bearing spikelets nearly to their bases. The plants furnish good grazing when young, and are sometimes cut for hay. River flats, especially where overflows occur. July-September.

89. Heleochloa
Spikelets 1 flowered, flattened, disarticulating above the glumes; floret awnless; lemma 1 nerved; caryopsis wall swelling and gelatinizing when wet. Fig. 292....................
.... *Heleochloa schoenoides* **(L.) Host**

Figure 292

Tufted spreading annual; culms 10-30 cm long; panicles short and thick, the base hidden in the upper sheath; spikelets about 3 mm long, flat. A minor weed of waste ground, railroad yards, etc. Northeastern and north central U.S.; California. Introduced from Europe.

90. Blepharoneuron
Inflorescence a panicle; spikelets small, 1 flowered, disarticulating above the glumes; lemma awnless, the 3 nerves hairy; palea hairy between the keels. Fig. 293...............
.....................**PINE DROPSEED**
Blepharoneuron tricholepis **(Torr.) Nash**

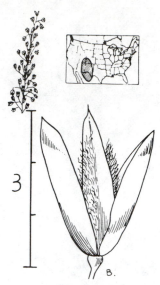

Figure 293

Figure 294

Perennial; tufted; culms slender, almost leafless, 20-60 cm tall. The leaf blades are crowded in a basal tuft about 1/3 as long as the culms. The panicles are loosely cylindrical, somewhat grayish. This species is a good forage grass, especially in early season. Open parks and thin woods, ponderosa pine, spruce, and fir forests. July-October.

91. Calamovilfa

Rhizomatous perennial sand-binders; inflorescence a panicle; spikelets 1 flowered, disarticulating above the glumes; floret with a tuft of long straight hairs on the callus. This genus was formerly placed in the Agrostideae, but differs from other members of that group in numerous microscopic characters. Fig. 294
. . . . **SAND REEDGRASS**
Calamovilfa longifolia (**Hook.**) **Scribn.**

Perennial; culms 50-180 cm tall, in small tufts; plants producing long, tough, scaly rhizomes; panicles large, 15-35 cm long; lemmas bearing copious tufts of straight white hairs on the callus. The plants are coarse and tough, but make considerable amounts of winter feed and are sometimes cut for hay. Sandy soil, hills and plains; shores of Lake Huron and Lake Michigan. August-September.

Calamovilfa gigantea (Nutt.) Scribn. & Merr. is similar but larger, and has hairs on the backs of the lemmas. Sand dunes of the Great Plains and southwestern states.

Tribe 22. Chlorideae

92. Spartina

Perennials, usually with rhizomes; inflorescence a panicle of 1-sided spikes, the spikelets in 2 rows along the lower side of the rachis, crowded, very flat and keeled, 1 flowered, disarticulating below the unequal glumes.

1a Leaf blades flat when fresh, 4-25 mm wide 2

1b **Leaf blades rolled, 2 mm or less wide. Fig. 295** . . . .
. . . . SALT MARSH GRASS
Spartina patens **(Ait.) Muhl.**

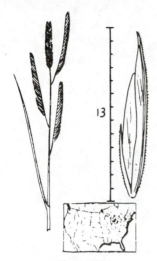

Figure 295

Perennial; spreading by slender, scaly rhizomes, forming large colonies in coastal salt marshes along the Atlantic and Gulf coasts. Culms up to 1 m tall, usually shorter. This grass is tough and harsh, but is frequently harvested for hay, which is used as packing material. The plants are valuable as landbuilders along the coast, trapping and holding the tidal mud. July-September.

Spartina gracilis Trin. (ALKALI CORDGRASS) is similar, but is found on alkali or salty flats in the interior of the United States, from the Dakotas to Kansas, westward to Washington and California.

2a **Leaf margins rough to the touch; plants widespread. Fig. 296** . . . .
. . . . SLOUGH GRASS
Spartina pectinata **Link**

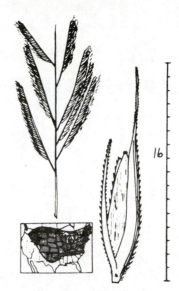

Figure 296

Perennial; spreading by tough, scaly rhizomes which are up to 1 cm thick; culms 1-2 m in height. The whole plant is coarse and tough, with saw-edged leaves. Slough grass was formerly one of the dominant grasses of the tall grass prairie region of the north central United States. Now it survives largely along roadsides, in ditches, and on wet ground. July-September.

Spartina cynosuroides (L.) Roth is taller, has numerous spikes; second glume without an awn. Salt marshes, Atlantic and Gulf coasts.

2b **Leaf margins smooth; plants of Atlantic coastal salt marshes. Fig. 297** . . . .
. . . . *Spartina alterniflora* **Loisel**

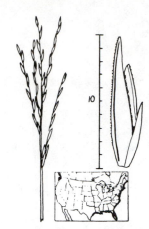

Figure 297

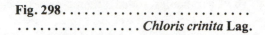

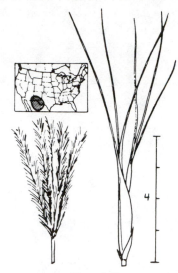

Figure 298

Perennial; spreading by rhizomes; culms .5-2.5 m tall, thick and spongy; leaf blades flat, 5-15 mm wide, tapering to a fine point; panicles narrow, the spikes slender and erect, 5-15 cm long; spikelets about 1 cm long, the floret smooth or slightly hairy. Coastal salt marshes, often growing in shallow water. Introduced along the coast of southwestern Washington. July-October.

Over a century ago, this species was introduced into Southampton, England in ships' ballast. There it hybridized with the English *S. maritima.* The hybrid, *S.* X *townsendii,* was sterile but later gave rise, by chromosome doubling, to *S.* X *anglica* Hubbard, a very successful colonizer of coastal mud flats.

93. Chloris FINGERGRASS

Annual or perennial grasses; tufted or with short stolons; inflorescences of 1-several whorls of slender, 1-sided spikes; spikelets sessile in 2 rows along the lower sides of a triangular rachis; disarticulation above the narrow, acuminate glumes; fertile floret 1 or rarely several, 3 nerved, awned; rachilla bearing 1-several modified rudimentary florets above the fertile one.

1a **Lemmas bearing 3 long awns, each longer than the body of the lemma.**

Perennial; tufted; plants 40-100 cm tall; leaf blades 2-4 mm wide. The silvery panicle of spikes is vase-shaped, 5-15 cm long, feathery because of the numerous long awns. Individual spikes are 5-10 cm long. Spikelets disarticulating above the glumes; second floret rudimentary, reduced to awns; lemmas 3 nerved, all of the nerves extending into the awns, which are about 1 cm long. The plants are sometimes cultivated for ornament. Fields and rocky open ground. May-September.

In American publications, this species has usually been called *Trichloris crinita.*

1b **Lemmas with a single awn** 2

2a **Lemmas without tufts of white hairs near the apex.** 3

2b **Lemmas bearing tufts of long, whitish hairs on the edges near the apex. Fig. 299** . . . .
. . . . **FEATHER FINGERGRASS**
Chloris virgata **Swartz**

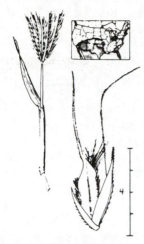

Figure 299

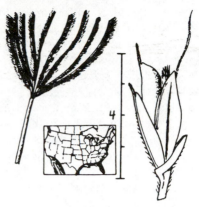

Figure 300

Annual; tufted; culms erect or spreading, 40-100 cm tall; some of the sheaths swollen; leaf blades 2-6 mm wide; spikes 2-8 cm long. The slender, vase-shaped panicles of spikes have a silky, white or pinkish cast because of the numerous long awns and fuzzy lemmas. The fertile lemma is about 3 mm long. The rudimentary floret is wedge-shaped and bears an awn about the same length as that of the fertile lemma. This species is found as a weed in fields, along roadsides and railroad tracks in the Southwest. The occurrences in the eastern states are probably introductions. In New England the plants occur on wool waste heaps around woolen mills, the seeds being imported in the raw fleeces. July-September.

3a Plants 1-1.5 m tall; leaf blades tapering to long, fine points; cultivated forage grass. Fig. 300. . . . . RHODES GRASS *Chloris gayana* **Kunth**

Perennial; spreading by leafy stolons; panicles vase-shaped, 5-10 cm long; spikelets yellowish, each with 2 rudimentary florets above the fertile one. Rhodes grass is grown in the southern states for hay and grazing and is found growing wild in fields and on waste ground. It winterkills at 20 degrees F., and so is adapted only to the far South. Introduced from Africa.

3b Plants 20-50 cm tall; leaf blades with blunt tips; native grass of central and southern U.S. Fig. 301. . . . . WINDMILL GRASS *Chloris verticillata* **Nutt.**

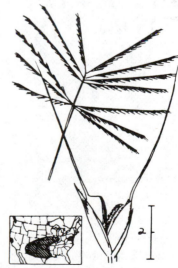

Figure 301

Perennial; culms tufted or with short stolons; leaf sheaths flattened and keeled; ligule membranous and ciliate, short; leaf blades grayish-green, 1-3 mm wide; panicles of 2 or 3 whorls of stiff, widely-spreading spikes, each 5-15 cm long. Spikelets rather widely spaced, the spikes slender; spikelets around 3 mm long; awns 5-8 mm long; fertile lemma hairy on the nerves; rudimentary lemma blunt. Windmill grass is primarily a plant of the plains of the Southwest, but may appear occasionally as a waif in the North. The mature panicles break off and roll as tumbleweeds. June-September.

94. Eustachys

Perennial grasses, tufted or with short stolons; leaf sheaths strongly keeled; inflorescence of several whorled 1-sided spikes; spikelets in 2 rows on the lower sides of a slender triangular rachis, placed at right angles to it; disarticulation above the glumes; second glume split at the tip, with a short awn tip; fertile lemma dark brown, nearly awnless, blunt. Fig. 302 . . . .
. . . . *Eustachys petraea* (Sw.) **Desv.**

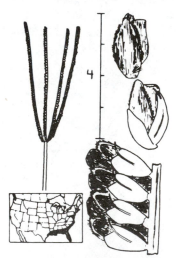

Figure 302

Perennial; tufted or with short stolons; plants up to 100 cm tall; spikes 4-10 cm long; lemmas chocolate brown; leaf sheaths and culms strongly keeled; leaf blades pale and with

rounded tips. Pine woods of the southern Atlantic and Gulf Coastal Plains. August-May.

Eustachys glauca Chapm. is similar but has inflorescence usually composed of 10-20 spikes. Coastal lowlands, North Carolina to Florida.

95. Gymnopogon

Tufted perennials with short, blunt leaves; inflorescence a panicle of very slender 1-sided spikes, borne singly at the nodes of the rachis; spikelets in 2 rows on the lower side of the rachis, usually 1 flowered, the rachilla extending beyond the fertile floret and bearing an awned rudimentary floret. Fig. 303 . . . .
. . . . . *Gymnopogon ambiguus* (Michx.) **B.S.P.**

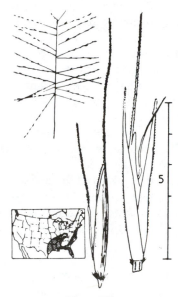

Figure 303

Perennial; tufted or with short rhizomes; culms 30-60 cm tall, stiff and erect, with overlapping sheaths and short, broad, spreading leaf blades. The slender spikes may reach 15-20 cm in length, and the whole panicle may be half the total height of the plant. Pine woods, mostly on the Atlantic and Gulf Coastal Plains. September-November, also in the spring.

The following species are similar but have short awns, less than 3 mm long:

G. brevifolius Trin. has a single fertile floret, and the branches lack spikelets toward the base. Coastal plains, from New Jersey to Florida, Arkansas, and Louisiana.

G. chapmanianus Hitchc. and *G. floridanus* Swallen. Both have 2-3 fertile florets and the branches bear spikelets to their bases. These species intergrade and can be distinguished only arbitrarily. Sandy pine barrens, Florida.

96. Ctenium

Tussock-forming perennials; culms tall, unbranched; inflorescence a single curved, 1-sided spike; spikelets densely crowded in 2 rows along the lower side of the rachis, disarticulating above the glumes; second glume bearing a protruding awn at the middle of the back; florets several, only the basal one fertile. Fig. 304 . . . .
. . . . **TOOTHACHE GRASS**
Ctenium aromaticum (Walt.) Wood

Figure 304

Perennial; tufted; plants 1-1.5 m tall. The bases of the plants are surrounded by the coarse, fibrous remains of the old sheaths. Toothache grass is a plant of wet pine woods on the sandy coastal plain. The fresh roots are said to have a spicy smell. The plants furnish some forage for cattle in the South. May-July.

97. Schedonnardus

Dwarf tufted perennial; inflorescence skeleton-like, of solitary 1-sided slender spikes borne along a twisted rachis; spikelets borne in 2 rows along the lower sides of a triangular rachis, 1 flowered, disarticulating above the glumes. Fig. 305 . . . .
. . . . **TUMBLEGRASS**
Schedonnardus paniculatus (Nutt.) Trel.

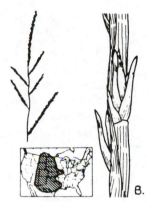

Figure 305

Perennial; tufted; plants 20-40 cm tall, with short, crowded basal leaves. Most of the height of the plant is made up of the skeleton-like panicle of very slender spikes. The main axis of the inflorescence is somewhat spiral. The whole panicle breaks from the plant when ripe and rolls with the wind, thereby distributing the seed. This is a grass of poor dry soil, especially on overgrazed or disturbed areas. Forage value very low. June-August.

98. Cynodon

Inflorescence a whorl of several 1-sided spikes; spikelets in 2 rows along the lower side of the rachis, 1 flowered, flattened, disarticulating above the glumes; lemma 3 nerved, awnless; rachilla extended behind the palea as a slender bristle. Fig. 306 . . . .
. . . . **BERMUDA GRASS**
Cynodon dactylon (L.) Pers.

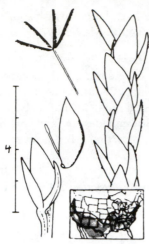

Figure 306

Perennial; producing both stolons and rhizomes and forming a fine green turf; erect portions of culms 10-40 cm tall, bearing about 4-5 slender spikes at the apex. Bermuda grass is one of the principal lawn grasses in the South and furnishes much pasturage as well. It is very persistent when once established and may become a serious weed pest on agricultural lands. Apparently native to the tropics of the Old World, but has become widely dispersed in warmer parts of the world. Blooming period extends through the warm season, and may be yearlong in the subtropics.

99. Bouteloua GRAMA GRASS

Perennials or annuals; tufted or with stolons or rhizomes; inflorescence of 1 many 1-sided spikes, borne in a raceme along an unbranched axis; spikelets crowded in 2 rows along the lower side of the rachis; fertile floret 1, the rachilla bearing a second awned rudimentary floret; disarticulation above the glumes except in a few species in which the whole spikes drop. Important western grazing grasses.

1a Inflorescence of 1-10 spikes, which remain on the plant; florets dropping from the glumes. 2

1b Inflorescence of numerous spikes, arranged in a slender raceme; entire spikes falling whole from the rachis when ripe. Fig. 307. . . . .
. . . . SIDE-OATS GRAMA
Bouteloua curtipendula (Michx.) Torr.

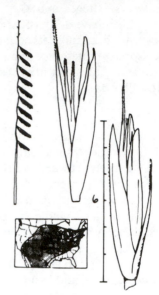

Figure 307

Perennial; culms in tufts, arising from slender rhizomes. Up to 80 cm in height, most of the height being the long raceme of drooping spikes. Late in the season all of the spikes drop from the flattened rachis, which remains, bearing only the short stalks of the spikes. Side-oats is a very attractive grass, with brilliant orange anthers, contrasting with the usually purple spikes. The name "side-oats" refers to the fact that most of the spikes droop toward one side of the rachis. One of the most valuable forage grasses in the western states, furnishing good feed yearlong. Dry plains and open rocky hillsides, from near sea level to 2700 m June-September.

2a Plants erect or spreading; internodes of the culms not woolly 3

2b Plant producing elongate stolons, the internodes covered with conspicuous short white wool. Fig. 308 . . . .
. . . . BLACK GRAMA
Bouteloua eriopoda Torr.

Figure 308

Perennial; plants sprawling, making bushy clumps, the stems thin and weak, rooting at the nodes. The 2-8 spikes are more slender than in the other species of this genus. Black grama is one of the best forage grasses in the southwestern states, furnishing good feed yearlong. Because of its ability to spread by stolons, it stands grazing well and recovers quickly from overuse. Open dry plains and hills, 670-1800 m elevation. July-September.

3a Rachis of the spike bearing spikelets to the tip; keels of glumes usually not black-dotted **4**

3b Rachis of the spike extending beyond the spikelets as a naked point 5-8 mm long; keel of the second glume bearing prominent black dots. Fig. 309 . . . .
. . . . HAIRY GRAMA
Bouteloua hirsuta Lag.

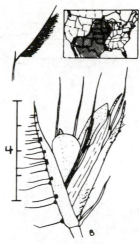

Figure 309

Perennial; tufted; culms 20-60 cm tall; leaves short, curly, mostly at the base of the plants. Hairy grama is a highly prized forage grass widespread west of the Mississippi, especially on dry plains and hills, from 670-1800 m elevation. In the Midwest it is usually found on dry hilltops. Hairy grama furnishes good feed yearlong, but is especially valued for winter forage. July-October.

4a Spikes 3-7 per culm, each 1-2 cm long; keels of glumes without long hairs; shallow-rooted annual plants with soft bases. Fig. 310 . . . .
. . . . SIX-WEEKS GRAMA
Bouteloua barbata Lag.

Figure 310

Annual; culms 10-30 cm long, in small tufts, erect or more commonly spreading out and forming flat mats. The seeds germinate after rains, and the plants mature and die rapidly, hence the name "six-weeks grama." The six-weeks grasses, belonging to a number of genera, furnish short-season feed after heavy rains, but their total forage production is small and they are highly undependable as forage plants. Dry plains, especially on overgrazed pastures. July- October.

4b Spikes 1-3 per culm, each 2.5-5 cm long; keels of second glumes bearing scattered long hairs; perennial, with hard bases. Fig. 311. . . . .
. . . . **BLUE GRAMA**
Bouteloua gracilis **(H.B.K.) Lag.**

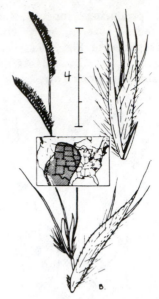

Figure 311

Perennial; tufted; culms 20-100 cm tall; most of the curly leaves are at the base of the plant. Blue grama is an excellent forage grass on the Great Plains and in the western mountains, furnishing good feed both summer and winter. Open plains, open or lightly timbered mountainsides. It may be confused with buffalo grass (Fig. 312), with which it often grows, but blue grama lacks the creeping stolons of buffalo grass. June-October.

100. Buchloë
Low creeping perennial; plants unisexual; pistillate spikelets enclosed in yellow, bony, bead-like structures borne in leaf axils; staminate spikelets borne on 1-3 1-sided spikes. Fig. 312 . . . .
. . . . **BUFFALO GRASS**
Buchloë dactyloides **(Nutt.) Engelm.**

Figure 312

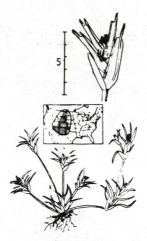

Figure 313

Perennial; spreading widely by stolons; pistillate spikelets enclosed in bead-like bodies with a short green crown on top; staminate spikelets in short, flag-like spikes. The two sexes are borne on separate plants. The diminutive plants are rarely more than 20 cm tall. Leaves short, curly, grayish-green. Despite its small size, this species is one of the most important forage grasses of the Great Plains, furnishing excellent forage yearlong. It closely resembles *Hilaria belangeri* (Fig. 314). For ways of distinguishing the two when not fruiting, see the discussion under that species. Buffalo grass frequently grows with blue grama grass (*Bouteloua gracilis,* Fig. 311) and resembles it, except that the grama grass lacks stolons. Blooming time mostly in the spring, but also later in the season.

101. Munroa
Low tufted sprawling annual; leaves short and stiff; inflorescence of reduced 1-sided spikes of a few spikelets, concealed within the sheaths of the upper leaves; florets several; lemmas 3 nerved, with a short, stiff awn; disarticulation above the glumes. Fig. 313 . . . . **FALSE BUFFALO GRASS** *Munroa squarrosa* (Nutt.) Torr.

Leaves of *Munroa* are stiff, harsh, and somewhat curled backwards. The spikelets are borne in 2's or 3's on short 1-sided spikes. The small vegetative leaves and the glumes are similar and hard to distinguish. On overgrazed or disturbed soil in blowouts, around prairie dog towns, corrals, etc. Forage value very low. June-August.

102. Hilaria
Rhizomatous or stoloniferous perennials of warm, dry climates; inflorescence a balanced spike; spikelets in groups of 3, which drop from the thin rachis as a group; lateral 2 spikelets 2 flowered, staminate; central spikelet of each group with 1 perfect flower.

1a **Plants spreading by slender creeping stolons; culms 10-30 cm tall. Fig. 314** . . . . **CURLY MESQUITE GRASS** *Hilaria belangeri* (Steud.) Nash

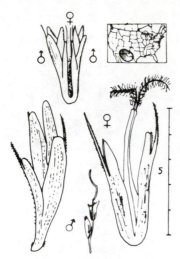

Figure 314

Hilaria jamesii (Torr.) **Benth.**

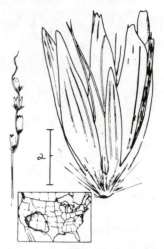

Figure 315

Perennial; forming extensive flat mats, the stolons rooting at the nodes. Dry plains, foothills, and brushy land, 650-1800 m elevation, often on heavy clay soils. Blooming occurs sporadically during the growing season.

Curly mesquite is an important range grass in the Southwest. Because of its stolons, it stands up well under heavy grazing. Where it occurs with buffalo grass, it is regarded as inferior to buffalo, being less productive and not curing as well. The plants greatly resemble those of buffalo grass (*Buchloë dactyloides,* Fig. 312), but may be distinguished by the spikelets and by the following features: 1. Color. Curly mesquite is light green when fresh, whitish when dry. Buffalo grass is grayish or olive green when fresh; tan, brownish or purplish when dry. 2. Stolons. The stolons of curly mesquite are round in cross section, very slender, and have tufts of hair at the joints. Those of buffalo grass are oval in cross section, stouter, and smooth at the joints.

H. swallenii Cory is similar, but the slightly larger glumes bear black dots. Texas and northern Mexico.

1b Plants erect, without stolons; rhizomes present. Fig. 315 ....

Perennial; culms tough, erect, 30-50 cm tall, from stout, scaly rhizomes. The racemes of clustered spikelets are whitish. After the groups of spikelets fall, the rachis remains behind as a thin, zigzag straw. Galleta (pronounced gieyetta) is an important forage grass on dry plains and deserts in the Southwest. It is fairly palatable to horses and cattle when fresh and green, but is scarcely eaten when dry. June-August.

Hilaria mutica Benth. (TOBOSA GRASS) is very similar, but the first glume on each lateral spikelet is fan-shaped. Heavy clay soils, especially on river bottoms where flooding occurs. Western Texas to Arizona.

Hilaria rigida (Thurb.) Benth. (BIG GALLETA) is larger, and has the culms covered with a dense white felt of hairs. Southwestern deserts.

Tribe 23. Pappophoreae

103. Cottea
Perennial, with hard knotty crowns; spikelets in panicles, disarticulating above the glumes and

between the florets; lemmas parallel-veined, their tips much lobed; nerves protruding as awns. Fig. 316 . . . .
. . . . *Cottea pappophoroides* **Kunth**

Figure 316

Foliage hairy; culms to 65 cm tall; panicles narrow, 10-18 cm long; spikelets bristly because of the many awns; marginal nerves of the lemmas separate nearly to the base; callus and lower margins of the lemmas copiously ciliate; palea ciliate on the lower margins; spikelets 7-9 mm long. Dry sandy slopes in canyons, between 600-900 m elevation; western Texas, Arizona, and Mexico. September-October.

104. Pappophorum
Tufted perennials; inflorescence a spikelike panicle; spikelets with 4-6 florets, but only the 1-3 lower ones fertile; disarticulation above the glumes; glumes equal, nearly as long as the entire spikelet; lemmas many-nerved, each nerve extending into an awn. Fig. 317 . . . .
. . . . **PAPPUS GRASS**
Pappophorum bicolor **Fourn.**

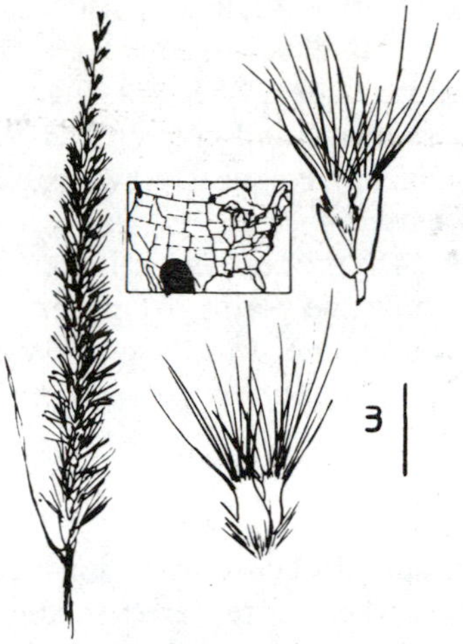

Figure 317

Erect, up to 80 cm tall; panicles straight and stiff, 10-20 cm long, bristly because of the many awns of the pinkish spikelets. Southwestern plains and deserts. April-November.

P. vaginatum Buckl. is similar but the spikelets lack pinkish coloration. Southern Texas to Arizona. April-November.

105. Enneapogon
Panicles gray, narrow and elongated; glumes as long as the spikelet; florets 3, only the basal one fertile, falling together; lemma many-nerved, with 9 awns, which are feathery-hairy on the lower half. Fig. 318 . . . .
. . . . **PAPPUS GRASS**
Enneapogon desvauxii **Beauv.**

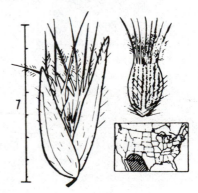

Figure 318

Perennial; tufted; culms 20-40 cm long; panicles spikelike. Leaf blades threadlike, about 1 mm wide. The lower sheaths are somewhat swollen, and contain cleistogamous spikelets which have nearly awnless lemmas. Several related species, in the genus *Pappophorum,* lack the feathery hairs on the awns. Dry desert hills and plains, from western Texas to Arizona and southward. Forage value low. Also called *Pappophorum wrightii.* July-November.

106. Orcuttia

Dwarf annuals; in dense tufts; inflorescence a short dense spike of a few erect spikelets which remain on the rachis after maturity; glumes shorter than the florets, the first narrowly lanceolate, the second usually 3 lobed; lemmas with many conspicuous parallel nerves, several extending into each of about 5-7 teeth at the broad apex; lower florets apparently staminate or sterile, the upper containing caryopses. Fig. 319 . . . .
. . . . *Orcuttia greenei* **Vasey**

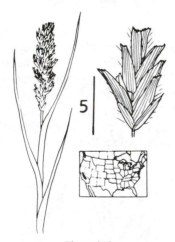

Figure 319

This small genus consists of 4 species native to drying pools and mud flats from Baja California to Shasta County, mostly in interior California. None is common. Spring.

Tribe 24. Zoysieae

107. Zoysia

Rhizomatous perennial grasses; inflorescence a slender raceme; spikelets laterally compressed, disarticulating whole, 1 flowered; first glume absent; second glume folded and keeled, the lower edges united; lemma hidden inside of the stiff glume. Fig. 320 . . . .
. . . . **ZOYSIA GRASS; MEYER ZOYSIA** *Zoysia matrella* **(L.) Merr.,** var. *japonica* **(Hack.) Forbes**

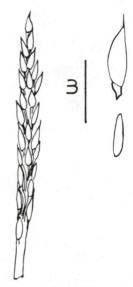

Figure 320

Leafy perennial; low; having many wiry rhizomes and forming a dense turf. The spikelets lack lodicules and do not open up. The elongated slender style branches emerge through the tip of the spikelet, followed later by the anthers. The glume has a split tip and a short awn.

This species is often grown for lawns in the southern states. It makes a good lawn, but turns brown and dormant early. Apparently, it seldom produces flowers, and is usually planted by "plugs," or small bits of turf. In the vegetative condition it may be recognized by its rather harsh, fine-tipped leaves and the thin, hard rhizomes.

108. Tragus

Tufted annual; spikelets borne in spiny clusters along an unbranched rachis; glumes covered with hooked prickles. Fig. 321 . . . .
. . . . **COCKLEBUR GRASS**
Tragus berteronianus **Schult.**

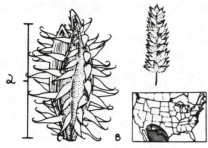

Figure 321

Annual; culms 10-40 cm long, spreading. The burs are borne along a slender raceme, from which they fall readily. Each bur consists of a group of 2-5 spikelets, but the second glumes of the 2 lower spikelets are covered with hooked prickles and conceal the remainder. On dry open ground in the Southwest and also at scattered points on the East Coast where wool is processed. The burs are readily transported by the wool of sheep. Probably introduced from the Old World; extending through the warmer portions of both hemispheres. August-October.

Subfamily VI. Panicoideae

Tribe 25. Paniceae

109. Melinis

Sprawling perennials; inflorescence a many-flowered panicle; spikelets disarticulating below the glumes; first glume minute, second glume and sterile lemma equal, concealing the floret; fertile floret 1, with a long awn. Fig. 322 . . . .
. . . . **MOLASSES GRASS**
Melinis minutiflora **Beauv.**

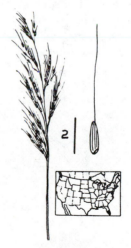

Figure 322

Molasses grass has a peculiar sweet odor. The foliage is covered with sticky hairs, a unique situation not found elsewhere among the grasses. The minute spikelets are strongly nerved. While of tropical African origin, the species is widely planted as a forage grass in the Americas and has been introduced into southern Florida. In Brazil it is reputed to repel mosquitoes because of the odor.

110. Panicum

Tufted or rhizomatous annual or perennial grasses; inflorescence a panicle, the branches sometimes nearly simple; spikelets dorsally compressed; first glume short; second glume and sterile lemma equal, papery, concealing the slightly shorter rigid fertile floret, which is shiny, awnless, and with inrolled edges which cover the keels of the flat palea. An enormous genus, most common in warmer climates of the world.

1a **Plants blooming twice, bearing panicles at the tips of the culms in spring or early summer and small axillary panicles later; winter rosettes of short, broad leaves present in most species; all perennials. Fig. 323. Note: the species of this group have been assigned to a new genus, _Dichanthelium,_ by Gould. The alternative names are indicated for the species given 2**

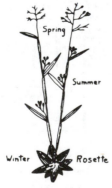

Figure 323

1b **Plants blooming once, all the panicles produced at the same period; winter roset-**
tes not present; plants annual or perennial 11

2a **Leaf blades of the culms less than 15 times longer than wide; plants forming winter rosettes of short, broad leaves 3**

2b **Leaf blades of the culms very narrow, 20 or more times longer than wide; plants without winter rosettes of broad leaves. Fig. 324 . . . . _Panicum depauperatum_ Muhl.**

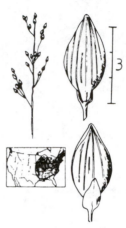

Figure 324

Perennial; tufted; plants 15-40 cm tall, with hairy or smooth leaf blades, up to 15 cm long and 2-5 mm wide. The terminal panicles produced in May and June, are open, pyramid-shaped, and on long, slender peduncles. The secondary panicles consist of a few spikelets and are concealed among the basal leaf blades. The second glume and sterile lemma form an empty beak which protrudes beyond the tip of the fertile lemma. Open dry woods and barren ground. May-June. _Dichanthelium depauperatum_ (Muhl.) Gould.

3a **Ligules conspicuous, of straight hairs, 2-5 mm long 4**

3b Ligules 1 mm or less long **5**

4a Upper leaf sheaths glabrous; lower
sheaths also usually glabrous; leaf blades
glabrous or sometimes hairy on the edges
only. Fig. 325 . . . .
. . . . *Panicum lindheimeri* Nash

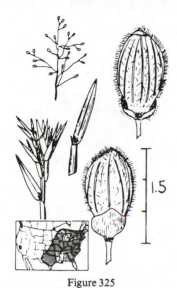

Figure 325

Perennial; tufted, at first rather slender,
30-100 cm tall. The plants later produce dense
tufts of short, leafy branches, in the axils of the
leaves, with small secondary panicles partially
concealed among these leaves. The culms may
then topple and the plants form flat circular
mats. Open dry ground and open dry woods.
Terminal panicles blooming from May-July.
Dichanthelium lindheimeri (Nash) Gould.

4b Leaf sheaths and usually the blades
conspicuously hairy. Fig. 326 . . . .
. . . . *Panicum lanuginosum* Ell.

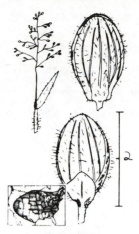

Figure 326

Perennial; tufted; plants 15-75 cm tall, with
open, pyramid-shaped terminal panicles
blooming in May and June. Later the plants
become much-branched, with loose axillary
tufts of short leafy branches, interspersed
with the short secondary panicles. Roadsides,
old fields, open woods, meadows, swamps.
Very common and widespread. *Panicum
lanuginosum* is usually broken up into a
number of scarcely distinguishable "species"
by other authors. *P. columbianum* (Fig. 333) is
quite similar. *Dichanthelium lanuginosum*
(Ell.) Gould.

5a Plants smooth or somewhat hairy, never
velvety to the touch **6**

5b Culms, leaf blades, and sheaths velvety to
the touch, grayish; a smooth, sticky ring
below each node. Fig. 327 . . . .
. . . . *Panicum scoparium* Lam.

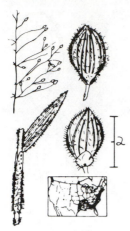

Figure 327

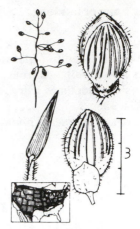

Figure 328

Perennial; tufted; plants 80-130 cm tall; leaf blades large, 12-20 cm long and 10-20 mm wide. The terminal panicles are open, elliptical, up to 15 cm long, produced in June and July. Later the plants become branched, with loose bunches of leaves in the axils of the sheaths of the main culm. The small secondary panicles are partially concealed among these branches. Low moist soil, mostly on the Atlantic Coastal Plain and northward in the Mississippi Valley. *Dichanthelium scoparium* (Lam.) Gould.

6a   Spikelets over 3 mm long 7

6b   Spikelets 2.7 mm or less long 8

7a   Upper leaf sheaths bristly hairy with spreading hairs; leaf blades 6-12 mm wide. Fig. 328 . . . .
. . . . *Panicum scribnerianum* Nash

Perennial; tufted; plants 20-50 cm tall, rather stiff; sheaths sparsely covered with stiff spreading bristles or nearly smooth. The terminal panicles, produced in May and June, are 4-8 cm long and about as broad, pyramid-shaped. After the terminal panicles have shed their spikelets, the plants become bushy-branched and produce small, simple panicles partially concealed by the tufted upper leaves. Prairies and open woods, often on dry sandy soil. *Dichanthelium oligosanthes,* var. *scribnerianum* (Nash) Gould.

7b   Upper sheaths glabrous or softly hairy; larger leaf blades 1.5-4 cm wide. Fig. 329 . . . .
. . . . *Panicum latifolium* L.

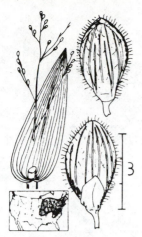

Figure 329

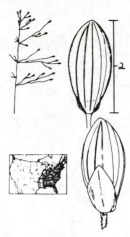

Figure 330

Perennial; tufted; 45-100 cm tall; terminal panicles produced in June, about 10 cm long and nearly as wide, with stiff spreading branches; spikelets hairy, 3.4-3.7 mm long. This is one of the most easily recognized of all grasses. The very broad leaf blades are heart-shaped at the base. Roadsides, woods, stream banks. The following species are also wide-leaved.

Panicum clandestinum L. has very bristly sheaths on the secondary branches. Spikelets 2.7-3 mm long.

Panicum boscii Poir. has spikelets over 4 mm long. The nodes of the stems are bearded with soft hairs, and the foliage is sometimes hairy.

8a Culms glabrous 9

8b Culms covered with short fuzz and sometimes with longer hairs 10

9a Spikelets glabrous, narrowly elliptical, 1.8-2.2 mm long; leaf blades not hairy on the margins. Fig. 330 . . . .
. . . . *Panicum dichotomum* L.

Perennial; tufted; plants erect, 30-50 cm tall, glabrous or with a ring of hairs on the lower nodes; terminal panicles in June, 4-9 cm long, with spreading branches. Later the culms become much-branched from the middle nodes, appearing like little trees and sometimes falling from their weight. The secondary panicles are small, with few spikelets. They extend slightly above the leaves of the branches. Widespread and common in rocky woods and on brushy land. *Dichanthelium dichotomum* (L.) Gould.

Panicum microcarpon Muhl. has strongly bearded nodes and tiny spikelets, 1.5-1.7 mm long. Moist woods, meadows, and swamps.

9b Spikelets minutely hairy, broadly obovoid at maturity, 1.5-1.8 mm long; leaf blades 7-14 mm wide, with long marginal hairs at the base. Fig. 331 . . . .
. . . . *Panicum sphaerocarpon* Ell.

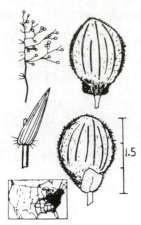

Figure 331

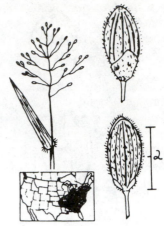

Figure 332

Perennial; tufted; culms 20-55 cm long, erect or spreading; plants glabrous except for the few hairs at the base of each leaf blade, and often somewhat glaucous. Leaf blades rather broad, the top one 4-9 mm in width. The broad panicle is less than twice as long as wide. Dry open ground and thin woods. Terminal panicles in June and July. *Dichanthelium sphaerocarpon* (Ell.) Gould.

   Panicum polyanthes Schult. has a longer, elliptical panicle, 2-4 times longer than wide, and an uppermost leaf 9-28 mm wide. Open woods and damp ground. Southern New England to Oklahoma and southward.

10a Spikelets 2.2-2.7 mm long; leaf sheaths glabrous; culm internodes covered with short, bent hairs. Fig. 332 . . . . *Panicum ashei* Pearson

Perennial; tufted; plants stiffly erect, 25-50 cm tall. The pyramid-shaped primary panicles are 5-8 cm long and have rather few spikelets. The internodes of the culms and portions of the leaf blades tend to be purple. The plants become rather sparingly branched, with elongated branches. Dry rocky woods, brushland, often on sandy ground. Primary panicles produced from May to July.

10b Spikelets 1.5-1.9 mm long; leaf sheaths hairy; at least some of the sheaths and internodes covered with a mixture of long hairs and short, fine fuzz. Fig. 333 . . . . *Panicum columbianum* Scribn.

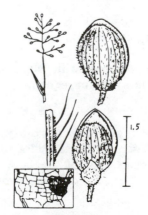

Figure 333

Perennial; tufted; plants 15-50 cm tall; leaf blades 3-6 cm long, 3-5 mm wide, the upper surface glabrous and the lower with fine, short hairs; primary panicles ovoid, 2-7 cm long, produced in June and July. Later the plants become much-branched and bushy, often making mats on the ground. Dry sandy or rocky sterile ground, in the open or in thin woods.

Panicum meridionale Ashe. Plants small and delicate; leaf blades 1.5-3 cm long, 2-4 mm wide; upper surfaces of leaves bearing long, erect hairs; spikelets 1.3-1.5 mm long. Sterile soil; Nova Scotia to Minnesota, south to Georgia and Alabama.

11a Spikelets glabrous or hairy, never warty **12**

11b Spikelets glabrous, covered with minute warts. Fig. 334. . . . .
. . . . ***Panicum verrucosum* Muhl.**

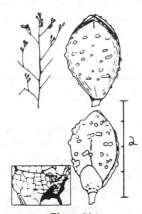

Figure 334

Annual; plants sprawling, the culms branching and rooting from the lower nodes, up to 150 cm long. The entire herbage is glabrous except for the margins of the sheaths; leaf blades thin, bright green, 5-20 cm long, 4-10 mm wide; ligules very short, hairy; panicles up to 30 cm long, very open; the small (about 2 mm long) spikelets borne in groups of 1-3 near the ends of the branches. Banks of streams; moist sandy or peaty soil. July-September.

Panicum brachyanthum Steud. has narrower leaf blades, 2-3 mm wide; pointed spikelets 3.2-3.6 mm long, covered with wart-based hairs. Arkansas and Oklahoma to Louisiana and Texas. August-September.

12a Fertile lemma minutely cross-wrinkled. Fig. 335. . . . .
. . . . **See genus 113. *Brachiaria***

Figure 335

12b Fertile lemma smooth and shining **13**

13a Plants without stolons; first glume much shorter than the whole spikelet. **14**

13b Plants producing long, wiry stolons; spikelets very blunt, the first glume nearly as long as the spikelet. Fig. 336 . . . .
. . . . **VINE MESQUITE GRASS**
***Panicum obtusum* H. B. K.**

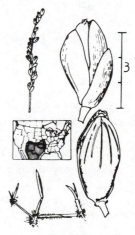

Figure 336

Perennial; producing elongated stolons up to 3 m or more in length, their nodes hairy, swollen; culms erect, flattened, 20-80 cm tall, with glabrous nodes; leaf blades 2-7 mm wide, glabrous; ligules 1 mm long, membranous; panicles 3-12 cm long, very narrow, with a few ascending branches; spikelets crowded, 3-3.8 mm long, glabrous, brownish; sterile lemma containing a palea and a staminate flower. Banks of streams, ditches, dry washes, irrigated fields. Vine mesquite furnishes some grazing, especially in the spring. Sometimes planted for erosion control, especially on terrace outlet channels, spillways of earth dams, and flood plain flats. June-September.

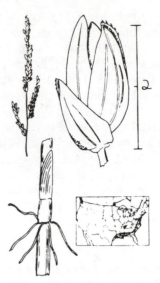

Figure 338

Perennial; culms hard and stiff, 50-150 cm tall; sheaths smooth or bristly; leaf blades 10-25 cm long, 7-15 mm wide, scabrous on top; panicles slender and spikelike, 15-30 cm long; spikelets 2.3-2.7 mm long. Wet ground and in water; ponds and ditches, wet fields, on the Atlantic and Gulf Coastal Plains. Sometimes maiden cane becomes a weed in wet fields. April-July.

14a Plants producing underground spikelets on root-like underground branches; aerial panicles sterile. Fig. 337. . . . . See genus 119. *Amphicarpum.*

Figure 337

14b Plants lacking underground spikelets; panicles fertile 15

15a Panicles with spreading or drooping branches; rhizomes present or absent. . 16

15b Panicles long and slender, with erect branches; plants producing extensive rhizomes. Fig. 338. . . . . MAIDEN CANE *Panicum hemitomon* Schult.

16a Panicle much-branched, open, the spikelets usually on long stalks, not confined to the lower sides of the branches; fertile lemma not hairy at the tip 17

16b Panicle with mostly unbranched main branches, the spikelets on short stalks, mostly on the lower sides of the branches; fertile lemma with a tuft of minute stiff hairs at the tip 20

17a Sheaths covered with stiff spreading hairs 18

17b Sheaths glabrous 19

18a Spikelets 3.5 mm long or shorter; panicle branches slender, stiff. Fig. 339 . . . .
. . WITCH GRASS *Panicum capillare* L.

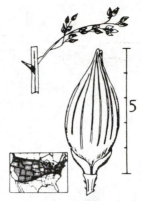

Figure 340

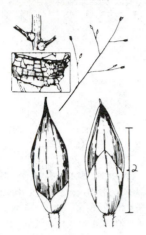

Figure 339

Annual; tufted; becoming bushy-branched, often 1 m or more tall; foliage soft, leaf blades hairy on both surfaces, 5-15 mm wide; terminal panicles large, dome-shaped, often more than half the length of the plant; numerous axillary panicles also present; bases of panicles usually hidden in the sheaths; axils of main panicle branches bearing tufts of hairs; panicles at maturity breaking away from the plant and rolling away as tumbleweeds.

One of the commonest weedy grasses of fields and disturbed soil; widespread in the United States. About 8 similar species are found in various parts of the country, but are much less common. July-October.

18b Spikelets 4.5 mm or more long; panicle branches stout, drooping. Fig. 340 . . . . .
. . . . BROOMCORN MILLET; PROSO MILLET *Panicum miliaceum* L.

Annual; tufted; plants 20-100 cm tall; foliage coarsely hairy or nearly smooth; leaf blades up to 30 cm long and 20 mm wide; panicles 10-30 cm long, drooping, the branches scabrous; spikelets 4.5-5 mm long, plump; fertile lemmas yellow, reddish, or brown. Proso is grown sparingly in the United States for forage, hog feed, and bird seed. The plants occur as strays on waste ground. Proso is probably native to Asia. Cultivated in the Orient, and to a lesser extent in Europe. It is supposed to be one of the most ancient of cultivated crops, and was known to the Romans under the name of *Milium,* whence comes our word millet. July-September.

19a First glume rounded or broadly triangular, 1/4-1/3 as long as the spikelet; plants annual, without rhizomes. Fig. 341 . . . .
. . . . FALL PANICUM *Panicum dichotomiflorum* Michx.

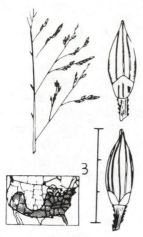

Figure 341

Figure 342

Annual; tufted; plants bushy, with freely branching erect or spreading culms; often coarse, with culms up to 2 m long in vigorous specimens; stems often zigzag, with an axillary panicle at each upper node. A common weed of cultivated fields, waste ground and moist soil around ponds or along streams. The size of the plants varies greatly, depending upon the moisture and fertility of the soil. July-October.

Panicum bartowense Scribn. & Merr. is similar but has bristly sheaths. Florida and the Antilles.

19b   First glume at least half the length of the spikelet, tapering to a sharp point; plants perennial, with hard, rhizome-producing bases. Fig. 342. . . . . SWITCH GRASS *Panicum virgatum* L.

Perennial; in clumps, spreading by thick scaly rhizomes; culms strong, 1-2 m tall; panicles large and open, 15-50 cm long; spikelets 3.5-5 mm long, often reddish and at times appearing laterally compressed. Switch grass is one of the most important native grasses of the tall grass prairie, but occurs far beyond the prairie area as well. Prairies, open ground, river banks and bottomlands, thin woods. A valuable forage species, sometimes harvested as a part of wild prairie hay. July-October.

20a   Plants producing rhizomes; spikelets scythe-shaped, set at an angle to the pedicel. Fig. 343 . . . . *Panicum anceps* **Michx.**

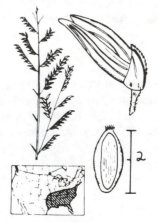

Figure 343

Perennial; short scaly rhizomes present; culms erect, 50-100 cm tall; leaves smooth or hairy, elongated, 4-12 mm wide; panicles open, 15-40 cm long; spikelets 3.4-3.8 mm long, curved. Open moist ground and woods, especially on sandy soil. July-September.

Panicum rhizomatum H. & C. is similar but has more elongated rhizomes, contracted panicles, and spikelets 2.4-2.8 mm long. Sandy soil on the Atlantic and Gulf Coastal Plains, Maryland to Texas; Tennessee.

20b Plants without rhizomes; spikelets straight or nearly so, set in line with the pedicel. Fig. 344 . . . .
. . . . *Panicum agrostoides* Spreng.

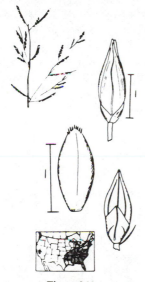

Figure 344

Perennial; tufted; culms 50-100 cm tall; panicles elliptical, the branches densely clustered with nearly sessile spikelets; spikelets green or somewhat reddish, 1.8-2.2 mm long. Moist shores and meadows, swamps, alluvial mud flats. July-September.

Panicum condensum Nash is similar but has a denser panicle; spikelets 2.2-2.5 mm long. Wet ground on the coastal plains; Pennsylvania to Florida and Texas.

111. Paspalidium

Somewhat succulent marsh or aquatic grasses; inflorescence a slender panicle made up of sessile, unbranched, 1-sided racemes that are erect and appressed to the rachis; spikelets borne in 2 rows along the lower 2 sides of a triangular rachis that ends in a naked point or a solitary spikelet. Spikelets dorsally compressed, awnless, acute, disarticulating entire; first glume broad, short; second glume shorter than the spikelet; lower lemma containing a large palea and a staminate flower; upper floret shorter than the lower one, its lemma rigid, minutely cross-wrinkled. This genus has spikelets similar to those of *Panicum,* but differs in the form of the inflorescence and the wrinkling of the fertile lemma. Figure 345 . . . .
. . . . . . *Paspalidium geminatum* (Forsk.) Stapf

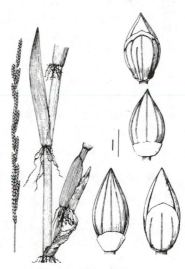

Figure 345

This species has thickish culms 40-140 cm long, often rooting at the lower nodes. The slender inflorescences, 10-30 cm long, are made up of numerous ascending short racemes. The spikelets are mostly 2.0-2.5 mm long, with a very blunt first glume. This species is of African origin, but occurs in the southern United States from Florida to Texas and Oklahoma. The plants grow on wet ground or in shallow water. Var. *paludivagum* (H. & C.) Gould has larger spikelets, 2.8-3.0 mm long. Florida, Texas; American tropics. The two lower spikelets in the illustration represent this variety.

112. Paspalum

Tufted or rhizomatous usually perennial grasses; inflorescence of 1 many 1-sided racemes, the spikelets borne in 2 or 4 rows along the undersides of the triangular or flattened rachis. Spikelets dorsally compressed, awnless, disarticulating below the glumes, flat on the sterile lemma side and convex on the second glume side; first glume usually absent or very small; second glume and sterile lemma of equal length and concealing the rigid fertile floret; fertile lemma with rolled-in margins that cover the edges of the flat palea of similar texture. Spikelets placed with the convex (fertile lemma) side facing toward the midrib of the rachis and the sterile lemma side facing outward. *Paspalum* is a very large genus of warm and tropical climates, a few of the species having forage value. See also the genera *Brachiaria, Eriochloa, Paspalidium* and *Axonopus,* all of which have somewhat similar spikelet arrangements.

1a Plants with creeping stems; aquatic or on wet ground 2

1b Stems not creeping. 3

2a Leaf blades, sheaths, and spikelets completely glabrous; spikelets about 2 mm long. Fig. 346. . . . . *Paspalum dissectum* L.

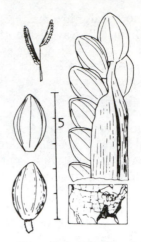

Figure 346

Perennial; plants creeping by extensive stolons; erect culms 20-60 cm tall, bearing 2-5 racemes, each 2-3 cm long; leaf blades dark green, 3-6 cm long, 4-5 mm wide. The rachis of the raceme is flat and thin, 2-3 mm wide. The first glume is lacking. Muddy flats and ditches and in shallow water. August-October.

2b Leaf sheaths with tufts of hairs on the auricles; second glume minutely hairy; spikelets 2.5-3.5 mm long. Fig. 347. . . . . KNOTGRASS *Paspalum distichum* L.

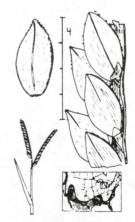

Figure 347

Perennial; spreading by long creeping stolons; erect culms 8-50 cm tall; inflorescence of 2 or rarely 3 racemes, each 2-7 cm long and somewhat curved, attached together at the summit of the culm; spikelets often with a minute first glume. Knotgrass forms large flat mats in ditches and on shores of rivers and ponds, usually in fresh water areas, rarely also in brackish localities. Found also in South America and Europe. May-September.

Paspalum vaginatum Swartz is similar but the spikelets are glabrous. Georgia, Florida, and Gulf Coast to Texas.

3a **Inflorescences borne at the tips of the culms and in the axils of the upper sheaths (sometimes concealed within the sheaths). Fig. 348. . . . . *Paspalum setaceum* Michx.**

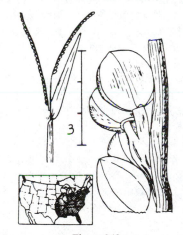

Figure 348

Perennial; culms tufted, arising from knotty crowns, up to 90 cm long, erect to prostrate; foliage glabrous to hairy; blades 2-20 mm wide. Racemes 1-5 on each peduncle, 3-17 cm long; spikelets paired, glabrous or pubescent, tan to purple, elliptic to circular, 1.4-2.7 mm long; first glume absent. Dry sandy land, open woods. Summer.

This species is extremely variable. A recent treatment by Banks includes 9 varieties, most of them formerly regarded as species. As a whole, the group ranges from New England to Minnesota, Nebraska, Texas, and the southeastern states.

3b **Inflorescences borne only at the tips of the culms, none in leaf axils or hidden in the sheaths. 4**

4a **Spikelets not fringed with long hairs . . . 5**

4b **Spikelets fringed with long silky hairs, borne on the edges of the second glume. Fig. 349. . . . . DALLIS GRASS**
Paspalum dilatatum **Poir.**

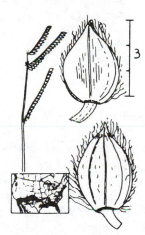

Figure 349

Perennial; tufted; plants 50-150 cm tall; inflorescence of 3-5 racemes, each 6-8 cm long; first glume absent. The spikelets are more pointed than those of most other species. Dallis grass is a valuable pasture grass in the southeastern states and under irrigation in the Southwest. The spikelets may become infected by an ergot fungus and hence become poisonous to cattle. Cultivated meadows and pastures and commonly escaped to the wild. Native of South America. May-September.

Paspalum urvillei Steud. (VASEY GRASS) is similar but has 12-20 racemes and strongly hispid lower sheaths. Virginia to Florida, Arkansas and Texas; California.

5a **Plants 1-2 m tall; spikelets 4-4.5 mm long. Fig. 350. . . . .**
 . . . . *Paspalum floridanum* Michx.

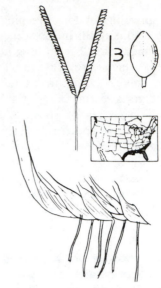

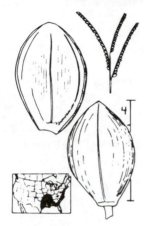

Figure 350

Perennial; culms single or in small tufts from short rhizomes; plants robust, 1-2 m tall; foliage hairy or glabrous; inflorescence of 2-5 racemes each 4-12 cm long. The axis of the racemes is strongly zigzag after the spikelets drop off. First glume absent. Low moist sandy barrens, Atlantic and Gulf Coastal Plains and northward in the interior. July-October.

Var. *glabratum* Vasey has glabrous foliage.

5b **Plants less than 1 m tall; spikelets less than 3.5 mm long. 6**

6a **Racemes 2, borne together at the end of the peduncle; stems unbranched, arising from short, woody rhizomes. Fig. 351. . . . .**
 . . . . BAHIA GRASS
 ***Paspalum notatum* Flügge**

Figure 351

Perennial; culms stiffly erect, with 2 spreading racemes; spikelets in 2 rows along each raceme, about 3-3.5 mm long. Bahia grass has been extensively planted for pasture and for erosion control along road shoulders in Florida and other southeastern states. One of the most conspicuous grasses along Florida roadsides. Introduced from tropical America. May-September.

6b **Racemes 3 or more, borne along a central rachis; stems sprawling or erect; plants lacking rhizomes 7**

7a **Spikelets mostly in 4 rows on each raceme, over 3 mm long; plants sprawling, often rooting at the lower nodes. Fig. 352. . . . .**
 . . *Paspalum pubiflorum* Rupr. ex Fourn.

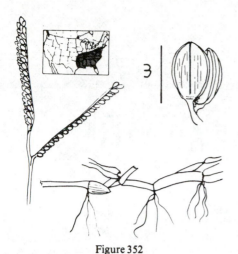

Figure 352

Culms up to 1 m long, spreading; leaf sheaths often bristly; blades flat, up to 15 mm wide; racemes 3 or more, up to 10 cm long, rather thick because of the 4 rows of spikelets; rachis flat, up to 2 mm wide; spikelets smooth or hairy, about 3 mm long. Moist open ground, shrubbery, stream banks. Autumn.

Var. *glabrum* Scribn. has glabrous spikelets.

7b **Spikelets in 2 rows on each raceme, less than 3 mm long; plants usually erect. Fig. 353.............................**
..............*Paspalum laeve* Michx.

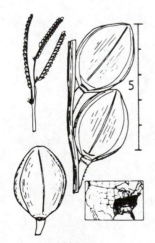

Figure 353

Perennial; tufted; plants 40-100 cm tall; foliage glabrous or hairy. The inflorescence consists of 3-4 racemes, each 3-10 cm long. The first glume is absent. A common species, varying greatly in hairiness of leaves and shape of spikelets. In typical plants, the spikelets are broadly oval. In var. *circulare* (Nash) Stone, they are nearly circular. Old fields, waste ground, meadow, open woods. July-October.

113. Brachiaria

Plants tufted or sprawling and rooting at lower nodes; inflorescence of few to many 1-sided racemes, the spikelets borne mostly in 2 rows on the lower sides of a triangular or flattened rachis, turned so that the first glumes are facing toward the midrib and the fertile lemmas are outward. Spikelets dorsally compressed, disarticulating below the glumes; first glume short to nearly as long as the spikelet; second glume and sterile lemma as long as the spikelet, the sterile lemma with a large palea and sometimes a staminate flower; fertile lemma rigid, minutely cross-wrinkled.

1a **Spikelets hairy.....................2**

1b **Spikelets glabrous 3**

2a **Spikelets 5-6 mm long; first glume not more than half as long as the spikelet. Fig. 354.........................**
.................TEXAS MILLET; COLORADO GRASS *Brachiaria texana* (Buckl.) Blake

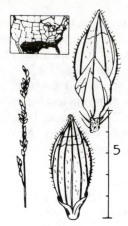

Figure 354

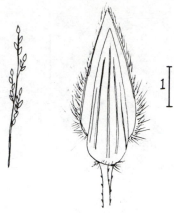

Figure 355

Annual; culms hairy, erect or decumbent and rooting at the lower nodes, usually 50-150 cm long; leaves 7-16 mm wide, 10-20 cm long, velvety; leaf sheaths velvety; ligule hairy, about 1 mm long; panicles narrowly cylindrical, 7-25 cm long, up to 3 cm in diameter; rachis and branches hairy. The lush, vigorous plants make good forage. Along streams and in corn and cotton fields. The common name, Colorado grass, apparently refers to the Colorado River of Texas, since this species does not occur in the state of Colorado. June-September. (*Panicum texanum* Buckl.).

Brachiaria arizonica (Scribn. & Merr.) Blake has similar spikelets, 3.5-3.8 mm long; blades 6-12 mm wide; panicle more open, with hairy branches. Western Texas to California and Mexico. August-September.

2b **Spikelets 3.5-4.5 mm long; first glume 3/4 as long as the spikelet. Fig. 355 . . . .** *Brachiaria ciliatissima* **(Buckl.) Chase**

Widely creeping perennial; leaf blades short, up to 8 cm long; foliage hairy; first glume glabrous; second glume and sterile lemma hairy, especially on the margins. The plants make flat mats on dry sandy ground, the prostrate culms rooting at the nodes. Texas, Oklahoma, and Arkansas. April-June, sometimes later.

3a **Racemes 2-6; spikelets 4.0-4.5 mm long, greenish; rachis 1-2 mm wide, flat. Fig. 356 . . . .** *Brachiaria platyphylla* **(Griseb.) Nash**

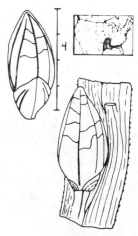

Figure 356

Annual; plants stooling out and sometimes rooting at the lower nodes; culms 25-40 cm or more long; leaf blades thickish, 4-12 cm long, 6-12 mm wide; rachis of the racemes flattened, 1-2 mm wide; spikelets glabrous, 4-4.5 mm long, the second glume and sterile lemma prolonged beyond the end of the fertile floret, forming a soft beak. Moist sandy ground. Formerly known as *B. extensa*. Summer.

3b **Racemes numerous; spikelets less than 3.5 mm long, yellow, brown, or deep purple; rachis triangular. Fig. 357** . . . .
. . . . **BROWNTOP MILLET**
Brachiaria fasciculata (Swartz) Parodi

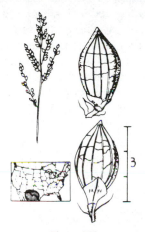

Figure 357

Annual; tufted, rather bushy; culms 30-100 cm long; leaf blades 6-20 mm wide, glabrous; ligule of hairs, 1 mm long; sheaths glabrous or papillose-hairy; panicles 5-15 cm long, made up of simple branches 5-10 cm long, the spikelets nearly sessile on the branches; spikelets frequently yellowish or brown, 2-3 mm long, with pronounced cross-veins between the longitudinal ones. Common weeds in fields, on river flats, and on waste ground. May-September. (*Panicum fasciculatum* Swartz).

114. Eriochloa
Inflorescence a panicle of short, 1-sided spike-like racemes; spikelets dorsally compressed,

disarticulating below the glumes, with a projecting cup-like structure at the base; second glume and sterile lemma pointed, concealing the floret. Fig. 358 . . . .
. . . . **PRAIRIE CUP GRASS**
Eriochloa contracta Hitchc.

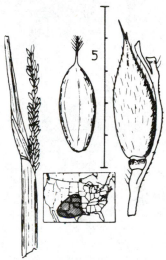

Figure 358

Annual; tufted, culms sometimes decumbent; plants 30-70 cm tall. The panicles are slender, consisting of nearly erect racemes. The cup-like swelling at the base of the spikelet is a modified first glume. The fertile floret is somewhat shorter than the sterile lemma and bears a short, hairy awn which is concealed within the spikelet. Open ground, moist places, ditches. June-October.

There are seven other species of this genus, all rather similar, in the southern states.

115. Anthaenantia
Perennials with rhizomes; inflorescence a panicle; spikelets dorsally compressed; first glume absent; second glume and sterile lemma equal, 5 nerved, covered with spreading hairs; sterile lemma with a thin, elongated membranous palea and often 3 stamens; fertile lemma and palea convex, brown, stiff, the lemma pointed, Fig. 359 . . . .
. . . . *Anthaenantia rufa* (Ell.) Schult.

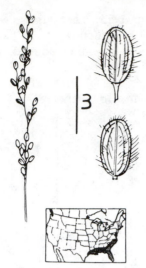

Figure 359

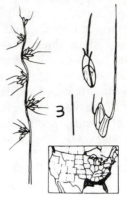

Figure 360

These grasses are important native forage plants on the Atlantic and Gulf Coastal Plains. They have unbranched culms with very elongated, blunt-tipped lower leaves. The upper leaves have very short blades. The spikelet hairs are up to 1 mm long and arise in dense rows between the nerves of the glume and sterile lemma. They are reddish or purplish in this species. July-November.

   Anthaenantia villosa (Michx.) Beauv. is similar but the spikelets have grayish hairs. The geographic range is about the same.

116. Oplismenus
Sprawling perennials, rooting at lower nodes; inflorescence a panicle of 1-sided racemes; spikelets dorsally compressed; first glume 3 nerved, shorter than the spikelet, with an awn 2-3 times its length; second glume 5 nerved, short awned; sterile lemma 7 nerved, slight¹ longer than the glumes, blunt and awnless; fertile floret stiff, shining, awnless, the margins of the lemma covering the edges of the palea. Fig. 360 . . . .
. . . . *Oplismenus setarius* (Lam.) R. & S.

Stems creeping; leaf blades short, ovate; panicles on long peduncles, small, made up of 3-5 racemes. Shady sites on the southern coastal plain along the Atlantic and Gulf. September-October.

   Variegated leaf forms of *O. hirtellus* (L.) Beauv. are in cultivation under the name of basket grass. The leaf blades have green, white, and purplish stripes.

117. Echinochloa
Tufted, often rather succulent, usually annual grasses; inflorescence a panicle of 1-sided, rather crowded racemes; spikelets disarticulating below the glumes, dorsally compressed, awned or sometimes merely pointed; first glume short; second glume and lower lemma pointed or awned, thin, covering the rigid, shining, perfect-flowered upper floret; lower lemma sometimes with a palea and a staminate flower; edges of fertile lemma rolled in near the base only.

1a Lower sheaths glabrous; fertile floret ovoid, 1.9-2.2 times longer than wide. Fig. 361 . . . .
. . . . **BARNYARD GRASS**
Echinochloa muricata (Beauv.) Fern.

Figure 361

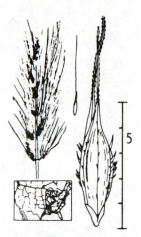

Figure 362

Annual; tufted; plants robust, up to 1.5 m tall; panicles up to 20 cm long, with spreading branches. The spikelets are covered with stout, spine-like hairs which arise from little yellowish blisters. Disturbed soil, in ditches, fields, marshes, borders of ponds. August-September.

Echinochloa crusgalli (L.) Beauv. is similar but does not have the blister-based stout spines on the spikelets. Just below the withering tip of the fertile lemma is a circle of minute hairs. These are absent in the preceding species. Fields and waste places; widespread; introduced from Europe.

1b   Lower sheaths bristly with stiff hairs; fertile floret elliptical, 2.5-3 times longer than wide. Fig. 362 . . . .
. . . . . *Echinochloa walteri* (Pursh) Heller

Annual; tufted; robust, up to 2 m tall; panicles up to 30 cm long. The spikelets usually bear awns 1-2.5 cm long. Individuals with glabrous sheaths can be identified by the narrow spikelets. Wet ground or shallow water, sometimes in brackish areas, mostly on the Atlantic and Gulf Coastal Plains. August-September.

118. Sacciolepis

Spikelets flat on the first glume side, bulged out at the base and much inflated on the second glume side; sterile lemma with a well-developed palea; fertile floret smooth and shining, dorsally compressed, much shorter than the second glume and sterile lemma. Fig. 363 . . . .
. . . . **SACK GRASS**
Sacciolepis striata (L.) Nash

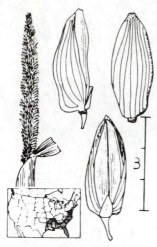

Figure 363

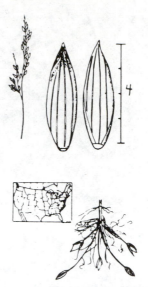

Figure 264

Perennial; culms often decumbent and rooting at the lower nodes, up to 2 m long. The leaf sheaths may be bristly-hairy or smooth. Panicles dense, cylindrical, 6-30 cm long. The spikelets are flat on the first glume side, very bulging at the base on the second glume side. Ditches, marshes and swamps, on the Atlantic and Gulf Coastal Plains. June-December.

Sacciolepis indica (L.) Chase is much smaller, with slender, wiry culms; panicle 1-4 cm long. Georgia; Texas. Spreading; native to Asia.

119. Amphicarpum
Inflorescence a slender panicle; spikelets usually lacking a first glume; second glume and sterile lemma longer than the floret and concealing it; floret hard, smooth, boat-shaped, the edges of the lemma covering the edges of the palea; aerial spikelets usually sterile; much enlarged fruitful underground spikelets borne on the tips of slender rhizomes. Fig. 364 . . . .
. . . . **PEANUT GRASS**
Amphicarpum purshii **Kunth**

Annual; plants hairy, tufted, erect, 30-80 cm tall. From the crown of the plant arise slender underground runners, 2-5 cm long, each bearing a single large spikelet, 7-8 mm long, at its tip. Most of the seed is produced by these underground spikelets, the aerial panicles being sterile. (See also Fig. 337, which shows a subterranean spikelet.) Sandy or peat soil, pine barrens of the Atlantic Coastal Plain. Fall.

Amphicarpum muhlenbergianum (Schult.) Hitchc. is perennial, has smooth leaves and stout underground rhizomes, bearing subterranean spikelets. Pine barrens, South Carolina to Florida.

120. Stenotaphrum
Perennial; spreading by coarse stolons; leaf sheaths strongly flattened and keeled; inflorescence an erect 1-sided spike, the axis thick, elliptical in cross section, the spikelets sunken into one side of it. Fig. 365 . . . .
. . . . **ST. AUGUSTINE GRASS**
Stenotaphrum secundatum (Walt.) **Kuntze**

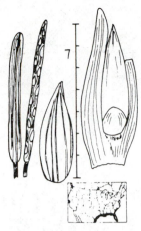

Figure 365

Perennial; plants spreading rapidly and extensively by stolons; erect flowering stems 10-30 cm tall. The leaf sheaths are very much flattened and keeled; blades with rounded, blunt tips. The spikelets are similar to those of *Panicum* species, but are nearly hidden by the flanges of the rachis. The first glume and sterile lemma are exposed. This unique grass is used for lawns in some localities in the South. Occasional plants with white-striped leaf blades may be cultivated for ornament. June-September.

Stolons of this species have alternating long and exceedingly short internodes, so that the leaves appear to be nearly opposite.

121. Setaria FOXTAIL

Panicles very narrow, cylindrical, usually bristly because of numerous bristles (sterile branches) interspersed with the spikelets; disarticulation just below the glumes, leaving the bristles on the rachis, except in millet and *S. magna*, which disarticulate above the glumes; first glume short, second glume and sterile lemma nearly as long as the rigid floret. A few species that have a bristle only below the terminal spikelet of each branch are now placed

in this genus, although they were formerly included in the genus *Panicum*.

1a **Bristles upwardly-barbed or smooth; panicles not clinging to objects** **2**

1b **Bristles downwardly-barbed, clinging to objects when the panicle is brushed upward. Fig. 366** . . . . **BRISTLY FOXTAIL** *Setaria verticillata* (L.) Beauv.

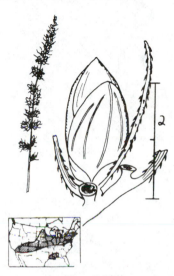

Figure 366

Annual; tufted; culms up to 1 m tall. The barbed bristles of the panicles not only cause them to adhere to wool, hair, or clothing, but also to each other. After windstorms, the plants will often be densely tangled. Sometimes flies or other insects are caught in the bristles. Common in cornfields and on disturbed soil. Introduced from Europe. June-October.

2a **Margins of sheaths bearing short hairs. . 3**

2b **Margins of sheaths smooth, thin and translucent; glabrous except for a few long hairs at the apex. Fig. 367 .... YELLOW FOXTAIL** *Setaria lutescens* **(Weigel) Hubb.**

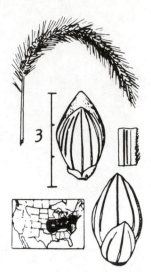

Figure 368

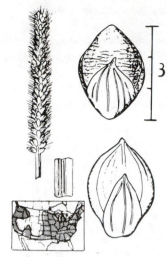

Figure 367

Annual; tufted; culms spreading or erect, up to 1 m tall; panicles stiff and compact, with a pronounced golden cast imparted by the yellow bristles; back of the fertile lemma exposed, transversely corrugated. An exceedingly common weed, in cornfields or other disturbed soil. Introduced from Europe. Also known as *S. glauca*. July-October.

Setaria geniculata (Lam.) Beauv. has similar panicles but is a perennial, and the culms arise singly or in small tufts from knotty, much-branched rhizomes. Atlantic and Gulf coast states, from Massachusetts to Texas, lower Mississippi Valley to Illinois and Iowa.

3a **Upper surfaces of leaves glabrous 4**

3b **Upper surfaces of leaves covered with soft hairs. Fig. 368 .... NODDING FOXTAIL; GIANT FOXTAIL** *Setaria faberi* **Herrm.**

Annual; tufted; culms up to 2.5 m tall; spikelets 2.6-2.9 mm long. This species looks much like green foxtail, but is usually larger and has more drooping, larger panicles. The velvety leaf blades are a good mark of recognition. Although known in North America for only about 40 years, nodding foxtail is already a bad weed in parts of the eastern and midwest states. Corn, soybean, and red clover fields; gardens; disturbed soil, especially on river bottomlands. Introduced from China. July-September.

4a **Spikelets dropping from the plants whole 6**

4b **Fertile floret when ripe "shelling out" of the glumes and sterile lemma, leaving them attached to the plant 5**

5a **Cultivated annual grain crop; panicles much lobed; fertile florets very plump, yellow, reddish, to black. Fig. 369......................... .... FOXTAIL MILLET** *Setaria italica* **(L.) Beauv.**

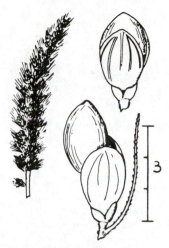

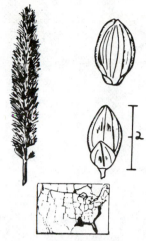

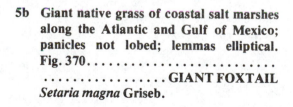

Figure 369

Figure 370

Annual; tufted; plants stout, often 1 m or more tall, with large, often definitely lobed panicles. Foxtail millet is regarded as being closely related to the wild green foxtail. The fertile lemma is variously yellow, orange, red, brown, or blackish. Millet is cultivated to a small extent as a forage and grain crop. Sometimes found persisting after cultivation or as a stray in the wild. Introduced from the Orient. July-October.

5b Giant native grass of coastal salt marshes along the Atlantic and Gulf of Mexico; panicles not lobed; lemmas elliptical. Fig. 370. . . . .
. . . . GIANT FOXTAIL
Setaria magna Griseb.

Annual; tufted; culms stout and tall, up to 2 cm thick and 4 m tall; leaf blades flat and scabrous, up to 50 cm long. The immense panicles reach lengths of 50 cm and diameters of 3.5 cm. They are somewhat nodding and may be lobed at the base, thickest at the middle and tapering toward the ends. Axillary panicles are much smaller than the terminal one. Bristles 1-2 cm long; spikelets about 2 mm long; fertile floret smooth and shining, brown when ripe. Giant foxtail is a characteristic plant of coastal marshes, found in the interior only in Arkansas. Its range also extends to the West Indies. August-September.

6a Plants annual, with soft bases and shallow roots; leaf blades usually less than 15 cm long, flat; panicles usually less than 10 cm long, with dense spreading bristles. Fig. 371. . . . .
. . . . GREEN FOXTAIL
Setaria viridis (L.) Beauv.

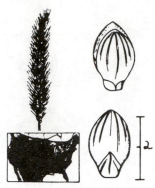

Figure 371

Annual; tufted; becoming much-branched from the base, 20-100 cm tall; leaf sheaths glabrous except for the short cilia along the margins; blades glabrous, usually less than 10 mm wide; panicles soft, slightly nodding near the tip; bristles green, rarely purple; spikelets 1.9-2.2 mm long, greenish except when ripe; second glume and sterile lemma nearly covering the fertile floret; fertile lemma nearly smooth. Green foxtail is one of the commonest weeds of cornfields and other areas of disturbed soil. The bristles are sterile panicle branches. Introduced from Europe. July-October.

6b Plants perennial, with hard, knotty crowns; leaf blades usually folded, 15-40 cm long; panicles slender, 2-25 cm long, with sparse bristles. Fig. 372 . . . . PLAINS FOXTAIL *Setaria macrostachya* **H. B. K.**

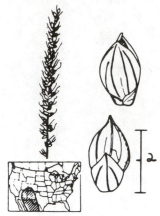

Figure 372

Perennial; in hard tufts; culms 40-120 cm tall; panicles slender, with the hairy rachis exposed between clusters of spikelets. The plants are leafy and are highly palatable to livestock, so that they are usually kept grazed down except in clumps of spiny bushes. Dry plains and savannas, especially along roadsides or other areas protected from grazing. April-October.

122. Pennisetum

Inflorescence a dense bristly cylindrical panicle; spikelets borne in small clusters, each surrounded by a group of long bristles (sterile branches), the group of spikelets and bristles falling from the axis as a unit; spikelets dorsally compressed; first glume short; second glume and sterile lemma enclosing the stiff smooth fertile floret. Fig. 373 . . . . **FOUNTAIN GRASS** *Pennisetum setaceum* **(Forsk.) Chiov.**

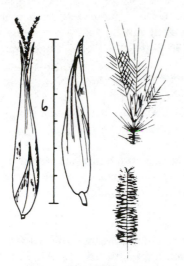

Figure 373

Perennial; tufted; culms up to 1 m tall, with a hairy, pinkish or lavender, spikelike panicle. The panicle is made up of a straight central axis, 15-35 cm long, bearing dense clusters of short branches. At maturity, each branch breaks off, carrying with it a spikelet or a small group of spikelets, surrounded by a dense tuft of long (3-4 cm) bristles. The larger bristles may have branches, feather-fashion. This is a handsome ornamental grass, introduced from Africa. Formerly known as *P. ruppelii*. Summer.

Pennisetum glaucum (L.) R. Br. (PEARL MILLET) has a thick spike, resembling that of the cat-tails. Rarely cultivated for forage in the South. Plants up to 4 m tall. Bristles of the spikelet-groups short, barely reaching the tips of the spikelets.

Nomenclatural arguments have resulted in this species being called *Pennisetum typhoideum* or *P. americanum* in publications.

123. Cenchrus SANDBUR

Tufted weedy plants; inflorescence a spike of spiny burs, each containing 1-several awnless spikelets; first glume short; second glume and sterile lemma mostly concealing the rigid floret; burs falling from the rachis when ripe, the spikelets remaining inside the bur and germinating there. Fig. 374 . . . . **FIELD SANDBUR** *Cenchrus longispinus* (Hack.) Fern.

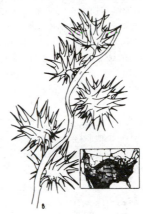

Figure 374

Annual; culms 20-90 cm long, usually spreading and making mats. The spikelets are mostly concealed by the horribly spiny burs, which are made up of sterile branches. The burs are borne in short spikes along a zigzag rachis, and fall off at a touch when ripe. The spines of the burs are very sharp, and each spine is microscopically backwardly-barbed. These spines can inflict painful and dangerous flesh wounds. Each bur contains 1 to several spikelets resembling those of species of *Panicum*. Sandbur is an undesirable weed of disturbed soil, much more common on sandy land than on heavier soils. Where it is abundant, it may furnish some feed for livestock when young. May-October.

Cenchrus tribuloides L. (DUNE SANDBUR) has larger burs, 10-17 mm in diameter. It grows on dunes along the Atlantic and Gulf Coasts.

Cenchrus incertus Curtis is a common species in the South. It has small burs with relatively few spines that are wide and flat at the base.

124. Digitaria CRABGRASS

Tufted or stoloniferous annuals or rarely perennials; inflorescence of several to many slender 1-sided racemes; spikelets borne singly, in pairs, or in trios, in 2 rows on the lower sides of a flattened or triangular rachis; spikelets disarticulating below the glumes, dorsally compressed, narrow and pointed; first glume minute or absent; second glume shorter than or equal to the sterile lemma; fertile floret narrowly ovate, pointed, the lemma firm, its edges flat, covering the margins of the equal palea.

1a Spikelets densely covered with long cottony hairs that conceal the bracts. Fig. 375.........................
.... COTTONTOP
Digitaria californica (Benth.) Henr.

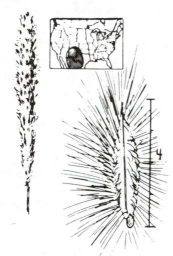

Figure 375

Perennial; tufted from knotty stooling crowns; plants 30-100 cm tall, with rather dense, slender panicles 5-15 cm long, silvery white or purplish; spikelets with a minute first glume; second glume shorter and narrower than the spikelet, densely silky; edges of the sterile lemma silky; fertile lemma brown, tapering to a point, its margins thin and white. Cottontop furnishes

good summer and winter feed in the Southwest, but is grazed mostly just after rains, when it makes rapid growth. Rocky ridges, margins of fields, in brush. July-November. Formerly called *Trichachne californica.*

Digitaria insularis (L.) Mez (SOURGRASS) is similar but the spikelet hairs are brownish. Florida to Arizona.

1b Spikelets not densely covered with long hairs, the surfaces of the glumes and sterile lemma plainly visible.......... 2

2a Rachis of the racemes thin and flat, with a pronounced midrib. Fig. 376......... 3

Figure 376

2b Rachis of the racemes triangular in cross section. Fig. 377 ....
.... *Digitaria filiformis* (L.) Koel.

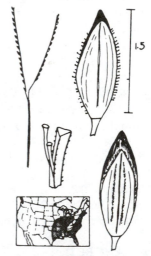

Figure 377

Annual; tufted; culms erect, 10-60 cm tall; racemes 1-5, up to 10 cm long. The first glume is absent, and the equal second glume and sterile lemma nearly cover the chocolate-brown fertile floret; spikelets about 1.5 mm long. Dry, usually sandy disturbed soil. August-October.

3a **Fertile floret brownish-black when ripe; leaf sheaths glabrous. Fig. 378. SMOOTH CRABGRASS**
Digitaria ischaemum **(Schreb.) Muhl.**

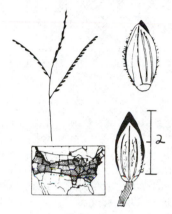

Figure 378

Annual; erect or spreading, often rooting at the lower nodes; culms usually 15-40 cm long; inflorescence usually of 2-6 racemes. This species frequently grows with the next and is a bad weed in lawns. Disturbed soil, fields, waste places, gardens, lawns. Introduced from the Old World. August-October.

3b **Fertile floret pale or leaden gray when ripe; leaf sheaths sparsely to densely covered with long straight hairs. Fig. 379. . . . . CRABGRASS**
Digitaria sanguinalis **(L.) Scop.**

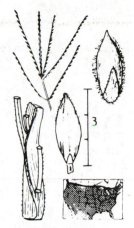

Figure 379

Annual; erect or spreading, usually rooting at the lower nodes; culms up to 1 m long, the plants often forming large mounds on rich soil. Crabgrass is a serious weed in lawns. Being originally from warm climates, it starts growth when hot weather arrives. The rampant plants soon make large patches in lawns, but die out after the first frosts. They also grow abundantly in fields and waste places, sometimes furnishing some forage. Introduced from the Old World. July-October.

125. Axonopus
Creeping perennial; inflorescence of several slender 1-sided racemes; spikelets dorsally compressed; first glume absent; second glume and sterile lemma covering the stiff elliptical floret. Fig. 380 . . . . CARPET GRASS
Axonopus affinis **Chase**

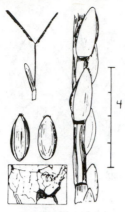

Figure 380

Perennial; producing extensive creeping stolons; erect culms flattened, 20-60 cm tall; racemes 2-5 on each peduncle, 3-10 cm long; rachis triangular in cross section, with the spikelets fitting closely against it. The first glume is absent. Low moist sandy or muck soil on the coastal plain, where it is important as a lawn and pasture grass. March-September.

Axonopus furcatus (Flügge) Hitchc. has glabrous spikelets, 4-5 mm long, the fertile floret shorter than the outer bracts. Moist coastal plain, Virginia to Florida and Texas.

126. Leptoloma

Perennial, tufted; panicle very open, the small spikelets on the tips of stiff, elongated slender pedicels; first glume minute or absent; second glume and sterile lemma concealing the pointed fertile floret; floret stiff, brown; edges of the lemma thin, covering the margins of the palea. Fig. 381 . . . .
. . . . **FALL WITCHGRASS**
Leptoloma cognata (Schult.) Chase

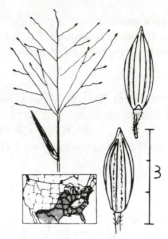

Figure 381

Perennial; plants stiffly spreading or erect, in large tufts; culms 30-70 cm long; panicles very open, with stiffly spreading, thin, somewhat zigzag branches. The pedicels of the individual spikelets may be up to 20-30 times the length of the spikelet. The second glume and sterile lemma bear appressed hairs. When mature, the panicles break off and roll away as tumbleweeds. Dry sandy open soil. May-September.

127. Rhynchelytrum

Panicles rosy-purple, turning gray; spikelets laterally compressed, concealed by the abundant hairs, disarticulating below the glumes; first glume minute, concealed by the hairs; second glume and sterile lemma equal, very hairy; fertile floret stiff, boat-shaped, acute. Fig. 382 . . . .
. . . . **NATAL GRASS; RUBY GRASS**
Rhynchelytrum repens (Willd.) Hubb.

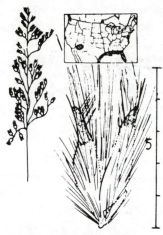

Figure 382

Figure 383

Perennial; tufted; plants about 1 m tall; panicles rosy-purple, 10-15 cm long; spikelets on bent or curled pedicels, and densely covered with purple hairs. The second glume and sterile lemma have short awns that are concealed by the hairs. The lateral compression of the spikelets is very atypical for the Panicoideae. Dry sandy land and open woods. Natal grass, an introduction from Africa, is grown in Florida for forage and has become naturalized there and also along the Gulf Coast. Formerly known as *Tricholaena rosea*. Winter.

Tribe 26. Andropogoneae

128. Imperata

Rhizomatous perennials; inflorescence spike-like, made up of racemes bearing pairs of equal, awnless, unequally-stalked spikelets, which disarticulate from the tips of the pedicels; glumes equal, concealing the shorter delicate sterile and fertile florets. Fig. 383 . . . .
. . . . **SATINTAIL**
Imperata brevifolia **Vasey**

Perennial; culms arising from hard scaly rhizomes; culms 1-1.5 m tall, with elongated leaves and slender, silvery-hairy panicles, 15-35 cm long and 1-3 cm thick. The spikelets are about 3 mm long and have a ring of long white hairs at the base, with some hairs also attached to the backs of the glumes. The spikelets fall from the rachis when ripe. Deserts. July-September.

Imperata cylindrica (L.) Beauv. (COGON GRASS) has been introduced in western Florida. Spikelets 4-5 mm long. This species has forage uses but may become a weed, because of its extensive rhizomes.

129. Erianthus

Tall perennial grasses; inflorescence plumy, of rames; rachis disarticulating, separating the spikelet pairs; spikelets of each pair equal, 1 sessile and the other stalked, with a tuft of long hairs attached to the callus; spikelets dorsally compressed, with long stiff glumes concealing the delicate sterile and fertile florets; awns usually present. Fig. 384 . . . .
. . . . **SILVER PLUMEGRASS**
Erianthus alopecuroides (L.) **Ell.**

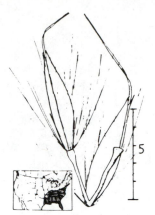

Figure 384

sterile and fertile florets; a tuft of silky hairs borne at the base of each spikelet and partly concealing it; disarticulation below each spikelet. Fig. 385 . . . . **EULALIA**
Miscanthus sinensis **Anderss.**

Figure 385

Perennial; culms strong and tall, 1-3 m long, arising from short, scaly rhizomes. The nodes, upper portions of the sheaths, and peduncles are appressed-hairy. The dense silky-hairy ellip- tical panicles are 20-30 cm long. Each yellowish spikelet bears a tuft of long silvery or purplish hairs from the base, as well as a few hairs on the upper portions of the glumes. Spikelets 5-6 mm long, with a flattened and twisted awn 1-1.5 cm long. The rachis breaks up into individual joints when ripe. Open woods, wet low ground, hammocks. September-October.

    Erianthus contortus Ell. is similar, but has dark brown glumes.

    Erianthus giganteus (Walt.) Muhl. (*E. saccharoides*) is similar but has straight, untwisted awns which are not flattened in cross section.

    Erianthus ravennae (L.) Beauv. (RAVENNA GRASS) is a cultivated perennial, with culms up to 4 m tall and large silky grayish plumes; it is hardy in the southern two-thirds of the United States. Awns very short or absent. Native to Europe. Fall.

130. Miscanthus
Tall perennials; inflorescence fan-shaped, of numerous long, silky racemes; spikelets paired, identical but on unequal stalks, dorsally com- pressed; glumes equal, concealing the delicate

Perennial, forming large clumps; culms 2-3 m tall, with plume-like, silvery-gray, fan-shaped panicles of long, hairy racemes. The spikelets are about 5 mm long, with a ring of hairs about as long as the spikelet, attached at the base of the glumes. The spikelets fall from the rachis when ripe. Cultivated widely as an ornamental, and occasionally escaping to the wild around inhabited places. Horticultural forms with white-striped or cross-banded leaves are also grown. Native to Asia. September-October.

    Miscanthus sacchariflorus (Maxim.) Hack., which has pure white panicles of awnless spikelets and spreads by vigorous rhizomes, has become a weed in the north cen- tral states and is cultivated for ornament. August-November.

131. Saccharum
Giant tropical grasses with large, plumy panicles; inflorescence made up of rames; spikelets paired, awned, equal, 1 sessile and the

other on a pedicel; rachis breaking up into individual joints when mature, each bearing a pair of spikelets; glumes equal, concealing the thin, delicate fertile and sterile florets. Fig. 386 . . . .
. . . . **SUGAR CANE**
Saccharum officinarum **L.**

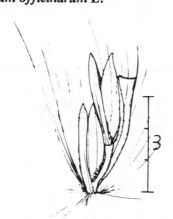

Figure 386

Perennial; tall stout plants, with culms 3-5 m tall and up to 3 cm thick. The stiff, elongated leaves have scabrous cutting edges. Panicles large and plume-like, 20-60 cm long. Sugar cane is widely cultivated in the tropics for the production of sugar, but in the United States is grown only in the southern end of the Mississippi Valley and in Florida. The plant seldom blooms.

   Erianthus ravennae (L.) Beauv. (RAVENNA GRASS) has a large, plume-like panicle, up to 60 cm long, and slender culms up to 4 m tall. Cultivated for ornament and hardy in the southern two-thirds of the country. It is similar to *Saccharum* in its nearly awnless spikelets, and has sometimes been put in this genus. See genus 129. *Erianthus.*

132. Schizachyrium
Tufted perennial grasses; sheaths often keeled; the individual inflorescence is a single rame on a bracted peduncle; culms usually branching and bearing numerous inflorescences toward the apex; rames disarticulating into individual internodes at maturity, each one bearing a sessile, awned, perfect-flowered spikelet and a reduced or abortive stalked one. Sessile spikelets stiff or leathery, the glumes of equal length and completely covering the internal bracts; first glume flattened, 2 keeled, the edges clasping the margins of the convex second glume; fertile lemma very delicate, bearing a twisted awn that is exserted at the tip of the spikelet. These species have usually been included in the genus *Andropogon,* and there is still controversy about their correct placement. Fig. 387 . . . .
. . . . **LITTLE BLUESTEM**
Schizachyrium scoparium **(Michx.) Nash**

Figure 387

Perennial; tufted; plants green or reddish, 50-150 cm tall; foliage smooth or hairy. The rames are borne on slender peduncles from the axils of the sheaths and the tips of the culms. The rachis joints and pedicels are strongly hairy. Little bluestem is characteristically a plant of dry prairies and plains, but occurs to some extent over nearly the entire United States. Prairies, old fields, rocky slopes and open woods. Little bluestem furnishes much grazing in the Midwest and West, especially for cattle and horses. August-October.

This species has been known as *Andropogon scoparius* in most American publications. A number of other species, similar in having a single rame on each peduncle, occur in the southern United States, but are much rarer.

133. Andropogon

Tufted or rhizomatous perennial grasses; inflorescence of several to many elongated rames arranged digitately or in a panicle; rames disarticulating into individual internodes at maturity, each one bearing a sessile, awned, perfect-flowered spikelet and a reduced or abortive stalked one. Sessile spikelets stiff or leathery, the glumes of equal length and completely covering the internal bracts; first glume flattened, 2 keeled, the edges clasping the margins of the convex second glume; fertile lemma very delicate, bearing a twisted awn that is exserted at the tip of the spikelet.

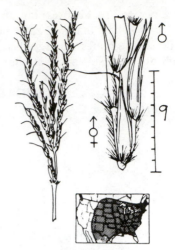

Figure 388

1a **Stalked spikelet sterile, reduced to a small rudiment or nearly absent; rames feathery, white** 2

1b **Stalked spikelet staminate, similar to the sessile one but awnless; rames green or purplish in color, not feathery with fine hairs. Fig. 388** . . . .
. . . . **BIG BLUESTEM**
Andropogon gerardii Vitman

Perennial; tufted or with short rhizomes; plants green or reddish, 1-2 m tall or even taller; foliage sometimes hairy. The plants bear 3-6 reddish rames at the tip of the culm, and usually some smaller inflorescences from the axils of the leaves. Big bluestem was one of the principal grasses of the tall grass prairie which produced the rich soils of our Cornbelt. Very little of such grassland still exists, but this species is still very common on untilled land in the prairie area. Farther east it is less common, but occurs on steep slopes, in meadows, and along river banks. This is an important forage species, and still provides much wild hay from native prairie. Also known as *A. furcatus* and *A. provincialis*. August-October.

Andropogon hallii Hack. (SAND BLUESTEM) has more elongated rhizomes and yellowish spikelets. Sandhills of the Great Plains and Rocky Mountain area.

2a **Inflorescence of 2-4 rames, which are enclosed at their bases by a leaf sheath. Fig. 389** . . . .
. . . . **BROOMSEDGE**
Andropogon virginicus L.

Andropogon saccharoides Sw.

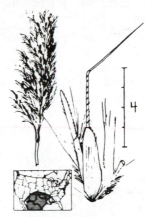

Figure 390

Figure 389

Perennial; tufted; 50-100 cm tall; foliage smooth or somewhat hairy, often reddish; culms bearing feathery-hairy inflorescences at the tip and from the axils of the leaves of the upper half of the culm. Broomsedge is a plant of sterile open hillsides, abandoned fields, and thin woods. It usually indicates poor soil. The forage value is apparently low, especially when the plants are mature. August-October.

About 14 other similar species or varieties occur in the southeastern states, mostly on the coastal plain. Most striking of these are the following:

Andropogon glomeratus (Walt.) B. S. P. (*A. virginicus,* var. *abbreviatus* (Hack.) Fern.). All the inflorescences are condensed into a dense, broomlike cluster at the top of the stem. Southeastern states; Texas to California, often on damp, low ground.

Andropogon elliottii Chapman has similar spikelets, but the reddish upper leaf sheaths are enlarged and cover most of the inflorescences. Atlantic Coastal Plain, New Jersey to Florida; lower Mississippi Valley to southern Illinois.

2b Inflorescence a terminal panicle of numerous rames; leafy bracts absent from

Perennial; tufted; rather bushy; culms 60-130 cm tall; foliage usually glabrous, becoming reddish when mature. The long-stalked panicles are silvery-white, oblong, 7-15 cm long. The rachis joints and pedicels are fringed with long white hairs. Sessile spikelet of each pair about 4 mm long, the pedicellate one rudimentary. Silver beardgrass and several of its close relatives are valuable forage grasses in parts of the Southwest, but are easily exterminated by overgrazing. Prairies and plains, rocky slopes, draws and dry washes, often on sandy soil. June-September.

Andropogon barbinodis Lag. is taller, with a short, fan-shaped panicle of spikelets 5-6 mm long. The nodes of the culms are prominently bearded. Southwestern United States and Mexico. Var. *perforatus* (Fourn.) Dewey is similar but has a circular pit in the center of the first glume of the sessile spikelet.

A number of similar species are found in the southwestern United States. Some authors separate them as the genus *Bothriochloa,* indicating that they differ in possessing rachis internodes and pedicels with a thin, grooved central line.

134. Elyonurus

Inflorescence an erect cylindrical rame, disarticulating into individual internodes, each bearing an awnless, nearly sessile spikelet and a stalked, staminate spikelet on a thick pedicel; spikelets dorsally compressed; first glume bearing 2 flanges on the back side which grasp the edges of the second glume; second glume slightly keeled, thinner than the first. Fig. 391 . . . . .
. . . . . *Elyonurus tripsacoides* **H. & B. ex Willd.**

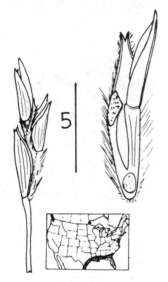

Figure 391

Erect plants, forming large clumps; perennial, with short rhizomes; culms 1 m or more tall, branching; rachis joints short and thick, disarticulating very obliquely; the staminate spikelet and its thick pedicel fitting closely against the "sessile" perfect-flowered spikelet. Ditches, roadsides, low ground, Gulf States and Mexico. June-September.

Elyonurus barbiculmis Hack. is similar but the spikelets are very woolly. Western Texas to Arizona.

135. Microstegium

Inflorescence of several rames; rachis internodes flat, widened toward the upper end; disarticulation at the base of each internode, which falls carrying at its base 1 stalked and 1 sessile spikelet, the 2 alike in size, both perfect-flowered, laterally compressed, and awnless. Fig. 392 . . . .
. . . . . . *Microstegium vimineum* (Trin.) Camus

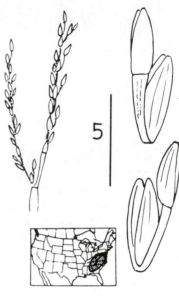

Figure 392

Straggling annuals, rooting at the lower nodes. The flat leaf blades are lanceolate. The equal glumes are papery in texture, unlike most members of this tribe. The inner bracts of the spikelet (sterile and fertile lemmas) are extremely small or missing altogether. If present, they are thin, nerveless scales. In var. *imberbe* (Nees) Honda, an exserted awn is present.

136. Arthraxon

Inflorescence a small, fan-shaped group of slender spikes; rachis disarticulating into individual internodes, each bearing a single spikelet at its base; spikelets lanceolate, pointed, laterally compressed; glumes equal, scabrous, greenish and papery. Fig. 393 . . . . . .
. . . . *Arthraxon hispidus* (Thunb.) Makino

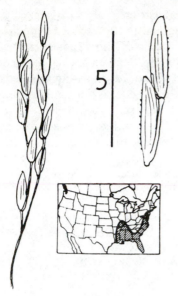

Figure 393

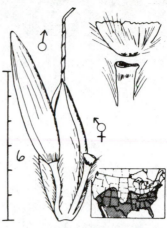

Figure 394

Low creeping annual with bristly sheaths and short broad blades with cordate bases. At the base of many spikelets a minute bristle is borne, representing the stalked spikelet usually found in this tribe. The inner parts of the spikelet are very delicate and membranous. The fertile lemma bears a minute awn, usually hidden inside the glumes. Low meadows, riverbanks; southern Pennsylvania to Florida, Louisiana, and Missouri. Introduced from the Orient. September-October.

137. Sorghum

Tall annuals or perennials; inflorescence a branched panicle of short rames consisting of 2-7 pairs of spikelets, all but the terminal segments bearing a hard, sessile, awned, perfect-flowered spikelet and a stalked, awnless, soft-textured staminate one; terminal segment with 2 staminate spikelets.

1a **Spikelets when ripe disarticulating from the tip of the pedicel; end of pedicel cup-shaped; rhizomes present; weed. Fig. 394 . . . .**

Perennial; culms tall and stout, 0.5-2 m tall, arising from thick, wide-spreading rhizomes; panicle open, pyramid-shaped, 15-50 cm long. The sessile spikelet is perfect-flowered and fertile, hard, and rather plump, about 5 mm long. The awn falls off readily, so the spikelets are often awnless. The pedicellate spikelets are of softer texture, staminate, narrower, and awnless. The sessile spikelet at the end of each short rame is accompanied by 2 pedicellate spikelets. The fertile spikelets vary from straw-colored to almost black. This species, regarded as a noxious weed in the southern states, is very similar to the annual crop, Sudan grass. Despite its bad traits, Johnson grass furnishes a great deal of forage and is readily eaten by livestock. Native to the Old World.

1b **Spikelets when ripe breaking from the plant with the upper end of the pedicel, leaving a jagged stub; rhizomes lacking; crop plant. Fig. 395 . . . .**
. . . . .SUDAN GRASS
Sorghum sudanense **(Piper) Stapf**

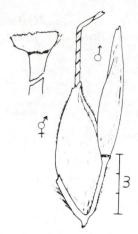

Figure 395

Annual; tufted; culms 1-3 m tall. The plants resemble those of Johnson grass but lack rhizomes. Widely cultivated for annual pasture and hay and sometimes found as a stray from dropped seed. Native to North Africa.

Sorghum bicolor (L.) Moench (SORGHUM). This species includes a large group of rather corn-like plants, cultivated in warm sections of the country for grain, fodder, silage, and syrup. The leaves and stems greatly resemble Indian corn, but the spikelets, similar to those of Johnson grass, are all borne in panicles at the tips of the stems. The grains may be black, brown, reddish, gray, or white. In many varieties, the grain becomes large enough to burst out of the glumes. Milo, hegari, feterita, durra, kafir corn, shallu, amber cane, broomcorn are all varieties of this species. Broomcorn is not used for forage, but the stiff, elongated panicle branches are the broomstraw of commerce.

The sorghums are of great value as forage crops, particularly in warm, dry areas. Plant breeders are constantly producing new cultivars for special purposes. Thus we now have dwarf, uniform hybrid strains that can be harvested with a grain combine, as well as tall, leafy forms, some with sweet juice, that are valued for forage and silage. Types with "waxy" grain can be used to produce "tapioca."

138. Sorghastrum

Tall perennials; inflorescence a much-branched panicle of stalked short rames, each consisting of a few rachis segments; each segment bearing a sessile, hard, dorsally-compressed, awned, perfect-flowered spikelet and a hairy sterile pedicel. The "stalked spikelet" is completely absent. The terminal segment of each rame bears 2 pedicels. Fig. 396 . . . .
. . . . **INDIAN GRASS**
Sorghastrum nutans (L.) Nash

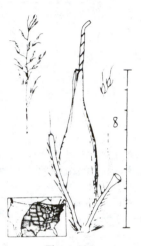

Figure 396

Perennial; in tufts from short rhizomes; plants 1-2.5 m tall; foliage usually smooth. Panicles narrow and rather dense, 15-30 cm long. The panicle has a "gold-and-silver" aspect because of the yellow, 5-6 mm long spikelets and the copious white hairs which fringe the rachises and pedicels. The prominent anthers are golden yellow. The panicle is made up of short rames of 1-3 joints. Each sessile spikelet is accompanied by a hairy pedicel. The rames break up into individual joints at maturity, each bearing a spikelet and a rachis joint and pedicel. Indian grass is one of the principal grasses of the tall grass prairie. It is also found in the eastern states and the Rocky Mountains. It forms an important component of wild prairie hay. Prairies, plains, stream banks, dry hills. July-September.

The following two species are similar, but have longer awns, 20-35 mm long, with 2 bends. Both occur in the southeastern states.

Sorghastrum elliotii (Mohr) Nash has dark brown spikelets and a loose, open panicle.

Sorghastrum secundum (Ell.) Nash has light brown spikelets; panicle 1-sided.

139. Coelorachis

Tufted perennials; inflorescence a single stiff cylindrical rame on each peduncle, disarticulating when mature into individual internodes, each of which bears an awnless, sessile, perfect-flowered spikelet and a thick-stalked rudimentary one that fits closely against the thickened rachis internode, the spikelets of the pair and the internode forming a cylindrical structure.

1a First glume with transverse corrugations. Fig. 397. . . . . *Coelorachis rugosa* (Nutt.) Nash

Figure 397

Perennial; culms from hard, knotty crowns; culms flattened, much-branched, 70-120 cm tall, with numerous axillary rames. Rames slender, "rat-tail"-like, brownish, 4-8 cm long, tapering from the middle toward the base and apex. They break up into individual joints readily, each joint bearing a perfect-flowered sessile spikelet and a sterile spikelet on a thickened pedicel. The first glume of the spikelet is strongly corrugated across the width. Wet pine woods and bogs, Atlantic and Gulf Coastal Plains. September. Formerly known as *Manisuris rugosa* (Nutt.) Kuntze

Other similar species are found in the southern states, differing in the degree of roughness, pitting, etc. of the first glume.

1b First glume smooth or pitted, not corrugated. Fig. 398 . . . . *Coelorachis cylindrica* (Michx.) Nash

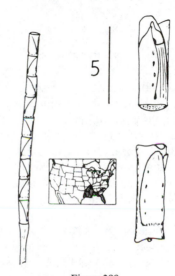

Figure 398

Culms erect, up to 1 m tall; inflorescences at the tip of the culm and in leaf axils, slender and curved, up to 15 cm long; spikelets 4-5 mm long, first glume pitted along the nerves. Pine woods and roadsides, Atlantic and Gulf Coastal Plains, northward to Missouri. September-October. Formerly known as *Manisuris cylindrica* (Michx.) Kuntze

140. Hackelochloa

Tufted, much-branched annual with terminal and axillary rames; spikelets paired, the sessile spikelet spherical, with a rough pitted blackish surface; rachis joint and pedicel united; stalked

spikelet flat, staminate. Fig. 399. . . . .
. . . . *Hackelochloa granularis* (L.) **Kuntze**

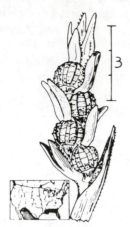

Figure 399

Annual; tufted; culms 30-100 cm tall, much-branched. The culms have numerous axillary rames; sheaths and culms covered with hairs which arise from little blisters. The individual spikelets are 1-2 mm long. The rames break up into individual joints, each bearing a sessile perfect-flowered spikelet which is blackish, and a strongly laterally compressed and winged staminate spikelet, which is green or reddish in color. This unusual grass is a native of the Old World tropics, but has been introduced into our southern states. It may yield some forage. Fall.

141. Eremochloa
Perennial, with many leafy stolons; inflorescence a single erect rame, elliptical in cross section, the overlapping spikelets all on one side; sessile spikelet dorsally compressed, the first glume with a membranous fringe at the apex, notched at the center; pedicellate spikelet minute, abortive, on a broad leathery pedicel. Fig. 400 . . . .
. . . . **CENTIPEDE GRASS**
Eremochloa ophiuroides (Munro) **Hack.**

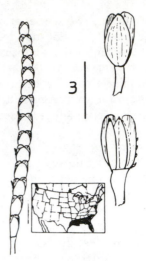

Figure 400

Centipede grass is cultivated in the southeastern states as a lawn grass. It makes a thick lush turf by the interweaving of many stiff stolons. The little purplish flowering spikes are usually 10-15 cm tall. The axis is rather wiry and does not seem to disarticulate. The sessile spikelets have a broad leathery first glume, with short bristles along the edges. Introduced from southeastern Asia. Summer.

142. Heteropogon
Inflorescence a single rame on a long peduncle; rachis when mature mostly disarticulating into single internodes, each with a sharp hairy point, bearing a pair of unlike spikelets. Fig. 401 . . . .
. . . . **TANGLEHEAD**
Heteropogon contortus (L.) **Beauv.**

Figure 401

Perennial; tufted; plants 20-80 cm tall; leaf sheaths flattened and keeled; rames borne at the tips of slender peduncles. The pairs of spikelets at the base of each rame are all staminate. In the upper portion of the rame, each pair consists of a sessile perfect-flowered spikelet and a stalked staminate spikelet. The sessile spikelet has a long, bent awn and a hairy rachis joint, which is attached below the base of the glumes. This spikelet greatly resembles the floret of some species of *Stipa*. Attached at the base of the perfect-flowered spikelet is a short pedicel bearing an awnless, laterally-compressed and winged staminate spikelet. The two spikelets of each pair fall as a unit. Because of the sharp hairy callus and stiff awn, the perfect-flowered spikelets may injure grazing sheep. A good forage grass when not in fruit. Rocky deserts in the Southwest; found throughout the tropics of both Old and New World. June-September.

Heteropogon melanocarpus (Ell.) Benth. has staminate spikelets with glabrous glumes, a row of glandular spots running down the middle of the glume. Southeastern states and Arizona.

143. Trachypogon

Inflorescence a single erect rame, its rachis not disarticulating; spikelet pairs consisting of a stalked, awned, perfect-flowered spikelet and a sessile, awnless staminate spikelet. Fig. 402 . . . .
. . . . **CRINKLE AWN**
Trachypogon secundus (Presl) Scribn.

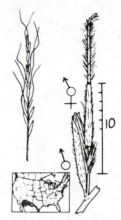

Figure 402

Perennial; tufted; 60-120 cm tall; herbage nearly smooth but the nodes bearing a circle of stiff, erect hairs. The erect slender rame is 10-20 cm long. The rachis remains whole with the short-pedicellate or sessile staminate spikelets attached to it. The longer-pedicellate perfect-flowered spikelets break from the rachis with their hairy, rigid pedicels attached. The perfect-flowered spikelet looks very much like the floret of some species of *Stipa*. The lemma is 6-8 mm long, with a bent and twisted awn 3-6 cm long. Rocky dry hills. May-October.

144. Coix

Corn-like annuals; pistillate spikelets borne in hard, bony beads, each borne on the apex of an axillary peduncle; staminate inflorescence of a few spikelets, borne on a stalk which emerges through the orifice of the bead. Fig. 403 . . . . . .
. . . . **JOB'S TEARS**
Coix lacryma-jobi L.

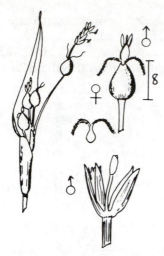

Figure 403

Plants coarse, up to 2 m tall, abundantly branching from the upper nodes. Each peduncle bears a hard, shiny, white, grayish, or black bead, from the upper end of which protrudes a short inflorescence consisting of a few joints bearing sessile and pedicellate staminate or sterile spikelets, in somewhat irregular combinations. Staminate spikelets consist of 2 glumes, enclosing 2 staminate florets. Within the bead is borne a single pistillate spikelet and 2 slender, tubular, sterile spikelets, along with the stalk of the staminate inflorescence. The stigmas protrude from the mouth of the bead. The staminate inflorescence breaks away and the beads fall from the plant when ripe. Job's tears is cultivated as a curiosity, and for the "beads," which are used in rosaries and jewelry. The plants may be found in the wild in the southern states. Introduced from the tropics of the Old World. Late summer.

145. Tripsacum

Peduncles at the tips of the culms and from the axils of the upper leaves, each bearing 1-several stiff spikes; spikelets unisexual; the bony lower segment of each spike bearing only pistillate spikelets, the thinner upper portion only paired staminate ones. Fig. 404. . . . . **GAMA GRASS**

Tripsacum dactyloides **L.**

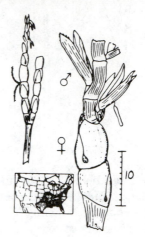

Figure 404

Perennial; in large clumps, from thick rhizomes. The plants reach 2-3 m in height. The spikes are born singly or 2-3 together at the tips of long leafless peduncles. The basal portion of each spike consists of a series of hardened, smooth, hollowed-out joints, each nearly enclosing a single pistillate spikelet. These joints break apart readily when ripe. The upper portion of each spike is made up of a series of joints, each bearing 2 sessile staminate spikelets. This portion of the spike is shed whole when the pistillate rachis breaks up. Gama grass is a close relative of corn and has been experimentally crossed with it. The plants are leafy and may produce some forage, but they are seldom abundant. River banks and moist ground. June-October.

146. Zea MAIZE, CORN

Tall, cultivated annual with broad leaf blades and a thick, solid culm; staminate spikelets paired, soft-textured, borne in a large terminal panicle (tassel) at the tip of the culm; pistillate spikelets borne in paired longitudinal rows on a thick axillary spike or cob that is enclosed by leafy sheaths (husks). Fig. 405 . . . . **MAIZE, INDIAN CORN, CORN** *Zea mays* L.

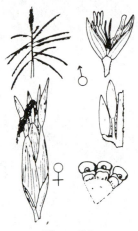

Figure 405

Tufted plants, exceedingly variable in size and habit, but characteristically with thick, solid stalks. The staminate inflorescence (tassel) is much-branched and bears pairs of staminate spikelets, one of each pair being sessile and the other pedicellate. Because of the crowding of the staminate spikelets, this arrangement may be obscured. The staminate spikelets have soft glumes and are 2 flowered. The pistillate inflorescence (ear) has paired rows of spikelets, so that the number of rows on a cob is usually even. The glumes, sterile and fertile lemmas form the "chaff" that remains on the cob when the kernels are removed. In a few cultivars, such as Country Gentleman, both florets of the pistillate spikelets develop grains, resulting in a very crowded ear, without the appearance of definite rows of kernels. Corn is unusual among grasses in having united styles, which form the "silk." Corn never persists after cultivation and is unknown in the wild state. It originated in tropical America and was exten-

sively selected and bred by the American Indians before the advent of Europeans in the Americas. All of the principal types of corn, such as dent, flint, sweet, flour, and pop corns, were produced by the Indian breeders.

The apparent ancestor of corn is a plant currently found in southern Mexico and northern Central America, and called by the populace "teosinte" or "madre de mais," the mother of corn. This plant, *Zea mexicana* (Schrad.) Reeves & Mangelsdorf, resembles corn in many vegetative features and in possessing a tassel of staminate spikelets. However, in place of ears, it possesses many small axillary spikes of hard, disarticulating internodes, each containing a single pistillate spikelet (Fig. 406).

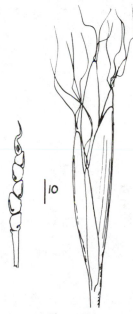

Figure 406

The pistillate spikes, although covered with husks as in corn, resemble the pistillate inflorescences of *Tripsacum*. Teosinte crosses readily with corn, and the offspring are intermediate in the structure of the pistillate inflorescences.

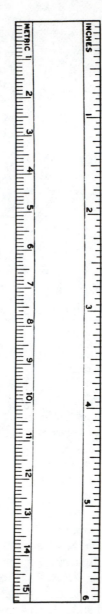

Figure 407

Index
and Pictured Glossary

APEX: The tip of a leaf blade, 4.
Fig. 408.

Figure 408.

AURICLE: A small, pointed or
rounded projection of the
sheath apex or base of the
blade, 4. Fig. 409.

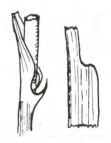

Figure 409.

AWN: A protruding midrib of a
glume or lemma, forming a
beard or bristle. Lateral
nerves rarely produce awns,
6. Fig. 410.

Figure 410.

B

BLADE: The elongated flat
spreading portion of a
grass leaf, 4. Fig. 411.

Figure 411.

Figure 412.

Figure 413.

Figure 414.

Figure 415.

F

FLORET: The unit of the spikelet, consisting of a flower and the lemma and palea that surround it, 5. Fig. 416.

Figure 416.

G

GLUME: An empty bract at the base of a spikelet. Usually 2 are present, 5. Fig. 417.

Figure 417.

H

I

INFLORESCENCE: A structure bearing spikelets (in grasses), 4.
INNOVATION: A leafy, non-flowering shoot arising at the base of a grass clump, 3.

J

K

L

LEMMA: The outer bract of a floret, which encloses the flower, 5. Fig. 418.

Figure 418.

LIGULE: A collarlike membraneous or hairy projection at the base of a leaf blade, 4. Fig. 419.

Figure 419.

LODICULE: A small blisterlike or flattened structure at the base of a grass flower that forces the floret open at blooming time, 5. Fig. 420.

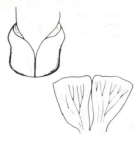

Figure 420.

Lolium, 27, 61
  perenne, 61
    var. italicum, 61
  temulentum, 61
Lovegrass, 117
Lycurus, 43, 136
  phleoides, 136

M

Madre de maís, 193
Maiden cane, 160
Maize, 192
Manisuris, 189
  cylindrica, 189
  rugosa, 189
Manna grass, 102
Mat muhly, 132
Meadow fescue, 57
Meadow foxtail, 88
Medusa-head, 92
Melica, 4, 30, 36, 55, 97
  bulbosa, 101
  fugax, 101
  imperfecta, 101
  mutica, 99
  nitens, 99
  porteri, 98
    var. laxa, 98
  smithii, 55, 99
  spectabilis, 100
  stricta, 98
  subulata, 100
  torreyana, 101
Meliceae, 97
Melinis, 35, 153
  minutiflora, 153
Meyer zoysia, 152
Microstegium, 43, 186
  vimineum, 186
    var. imberbe, 186
MIDRIB: The enlarged or conspicuous central strip of
  the leaf blade, containing
  one or more vascular
  bundles, 4.
Milium, 25, 89
  effusum, 89
Milo, 188
Miscanthus, 21, 35, 182
  sacchariflorus, 182
  sinensis, 182
Molasses grass, 153
Monerma cylindrica, 111
Monermeae, 110
MONOECIOUS: Having flowers
  of separate sexes, but both
  borne on the same plant.
  See also DIOECIOUS.

Mountain brome, 48
Muhlenbergia, 24, 26, 131
  arenacea, 132
  asperifolia, 132
  brachyphylla, 134
  bushii, 134
  cuspidata, 135
  emersleyi, 135
  frondosa, 133
  glomerata, 133
  mexicana, 134
  montana, 134
  porteri, 135
  racemosa, 133
  richardsonis, 132
  schreberi, 131
  squarrosa, 133
  sylvatica, 134
  torreyi, 136
  wrightii, 135
Munroa, 22, 26, 149
  squarrosa, 149
Mutton grass, 70

N

Natal grass, 180
Needle and thread, 108
Needlegrass, 114
Neeragrostis reptans, 118
New Zealand bent, 84
Nimble Will, 131
Nit grass, 86
NODE: The usually swollen joint
  of a stem, at which a leaf
  sheath is attached, 3.
  Fig. 421.

Figure 421.

Nodding fescue, 57
Nodding foxtail, 174

O

Oat grass, 112
Oats, 1, 7, 14, 74
Oniongrass, 100, 101
Oplismenus, 38, 170
  hirtellus, 170
  setarius, 170
Orchard grass, 14, 73
Orcuttia, 40, 152
  greenei, 152
Oryza, 23, 37, 45
  sativa, 45
Oryzeae, 13, 45
Oryzoideae, 13, 45
Oryzopsis, 24, 108
  asperifolia, 109
  hymenoides, 109
  micrantha, 109
  millacea, 110
  racemosa, 109

OVARY: The swollen lower portion of the pistil, containing the ovule or seed, 5.
  Fig. 422.

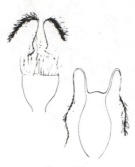

Figure 422.

P

PALEA: The inner of the two
  bracts that enclose a grass
  flower, 5. Fig. 423. See also
  LEMMA.

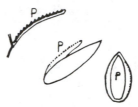

Figure 423.

Pampas grass, 112
Paniceae, 16, 153
PANICLE: A much-branched inflorescence bearing
  spikelets on pedicels, 5.
Panicoideae, 16, 153
Panicum, 1, 39, 89, 154, 173
  agrostoides, 163
  anceps, 162
  ashei, 158
  bartowense, 162
  boscii, 157
  brachyanthum, 159
  capillare, 161
  clandestinum, 157
  columbianum, 155, 158
  condensum, 163
  depauperatum, 154
  dichotomiflorum, 161
  dichotomum, 157
  fasciculatum, 169
  hemitomon, 160
  lanuginosum, 155
  latifolium, 156
  lindheimeri, 155
  meridionale, 159
  microcarpon, 157
  miliaceum, 161
  obtusum, 159
  polyanthes, 158

  rhizomatum, 163
  scoparium, 155
  scribnerianum, 156
  sphaerocarpon, 157
  texanum, 168
  verrucosum, 159
  virgatum, 162
Pappophoreae, 16, 150
Pappophorum, 29, 151
  bicolor, 151
  vaginatum, 151
  wrightii, 152
Pappus grass, 151
Parapholis, 41, 110
  incurva, 110
Paspalidium, 39, 163, 164
  geminatum, 163
    var. paludivagum, 164
Paspalum, 38, 164
  dilatatum, 165
  dissectum, 164
  distichum, 164
  floridanum, 166
    var. glabratum, 166
  laeve, 167
    var. circulare, 167
  notatum, 166
  pubiflorum, 166
    var. glabrum, 167
  setaceum, 165
  urvillei, 166
  vaginatum, 165
  virgatum, 165
Peanut grass, 172
Pearl millet, 177
PEDICEL: The stalk of a single
  spikelet, 5. Fig. 424.

Figure 424.

PEDUNCLE: The stalk of an inflorescence, 5.
Pennisetum, 20, 40, 176
  americanum, 177
  glaucum, 177
  ruppelii, 177
  setaceum, 176
  typhoideum, 177
Phalarideae, 15, 79
Phalaris, 24, 30, 80
  arundinacea, 81
  canariensis, 81
Phleum, 1, 25, 38, 88
  alpinum, 88
  pratense, 88
Phragmites, 22, 30, 111
  australis, 111
  communis, 112
Phyllostachys, 44
Pine dropseed, 139
Pinegrass, 58, 82
Piñon ricegrass, 110
Piptochaetium, 24, 110
  avenaceum, 110
  avenacioides, 110
  fimbriatum, 110

PROPHYLLUM: The membranous structure found between the main stem and the base of a branch, 3. Fig. 425.

Figure 425.

PSEUDOPETIOLE: A short stalk, between the apex of the sheath and the base of the leaf blade, occurring in bamboos.
PULVINUS: A swelling at the base of the branch of an inflorescence, or at the node of a culm, 5.

Q

R

RACEME: An unbranched inflorescence, bearing spikelets on pedicels that are attached to the central rachis, 5. Fig. 426.

Figure 426.

RACHILLA: The central stalk of a spikelet, to which the florets are attached, 5. Fig. 427.

Figure 427.

RAME: An unbranched inflorescence that bears some stalked and some sessile spikelets, usually in pairs or trios, as in *Andropogon* and *Hordeum*, 4.
RHIZOME: A creeping stem that grows at or below the soil surface and bears only scale leaves, 3. Fig. 428.

Figure 428.

S

SESSILE: Attached directly to an axis, lacking a stalk or pedicel, 5. Fig. 429.

Figure 429.

SHEATH: The tubular portion of the leaf, usually with overlapping edges, that surrounds the stem internode, 4. Fig. 430.

Figure 430.

Figure 431.

Figure 432.

Figure 433.